FLAME SPECTROSCOPY:
ATLAS OF SPECTRAL LINES

FLAME SPECTROSCOPY: ATLAS OF SPECTRAL LINES

M. L. Parsons and P. M. McElfresh

Department of Chemistry
Arizona State University
Tempe, Arizona

IFI/PLENUM · NEW YORK—WASHINGTON—LONDON · 1971

Library of Congress Catalog Card Number 76-165368

SBN 306-65156-4

© 1971 IFI / Plenum Publishing Corporation
A Subsidiary of Plenum Publishing Corporation
227 West 17th Street, New York, N. Y. 10011

United Kingdom edition published by Plenum Press, London
A Division of Plenum Publishing Company, Ltd.
Davis House (4th Floor), 8 Scrubs Lane, Harlesden, NW10 6SE, England

Printed in the United States of America

PREFACE

This atlas was begun mainly to gather together information on atomic absorption spectral lines for the use of practicing analytical chemists, who often find it necessary to use less sensitive lines. It was hoped that pertinent data could be obtained and for the first time published in a single format in one place.

This effort led to the realization that many workers in the field employ atomic emission and atomic absorption as complementary techniques. Therefore, it was decided to include both of these techniques in the atlas. Finally, it was decided that because atomic fluorescence spectroscopy shows so much promise as an analytical tool, the available data for this method should be included as well.

Since these three techniques provide fruitful research areas today, it is not possible to prepare a compilation of this scope and remain completely up to date. For practical reasons a cutoff date has to be set at which organization and typing begin. For this atlas, in most cases the literature references are complete through 1969. It is felt, however, that the absence of later references, especially in the areas of flame emission spectroscopy and atomic absorption spectroscopy, will not impair the usefulness of the atlas for the practicing analyst to any great degree.

ACKNOWLEDGMENTS

The authors are greatly indebted to Dr. J. D. Winefordner, who gathered together most of the information on atomic fluorescence spectroscopy, using a different format. The authors are also indebted to Mrs. Betty Bulechek, the typist.

CONTENTS

Section I

Atomic Absorption Spectral Lines

Section II

Atomic Emission Spectral Lines

Section III

Atomic Fluorescence Spectral Lines

SECTION I

ATOMIC ABSORPTION SPECTRAL LINES

Section I

ATLAS OF ATOMIC ABSORPTION SPECTRAL LINES

Introduction

To date no general atlas of atomic spectral lines which are
useful in atomic absorption spectroscopy has been published, although much experimental work has been done in this area. Even the
rather numerous books which have been recently published list only
a few spectral lines which are used in atomic absorption. Generally speaking, only the most sensitive atomic absorption line is used;
however, many more papers are appearing in which the concern is for
elements present in concentrations which are considerably greater
than trace. This is especially true in the areas of geochemical
and biological analyses. Advancements in instrument technology, in
stability of hollow cathode discharge lamps, and in lamp intensity
have permitted the analysis of major constitutents with much
greater precision. It is often convenient to avoid excessive di-
lution by utilizing lines of less sensitivity. Furthermore, there
are a number of elements for which the most sensitive line is in
doubt.

This atlas has been compiled so that the worker can determine
which lines should be used for specific concentration levels of
analyte. Margoshes[50] has given some very general guidelines for
utilizing the atomic transition probability for a particular line
in making the decision as to what line can be used. It was felt,
however, that the experimentally obtained data put on a uniform
basis would provide the researcher with a more realistic guideline
as to what lines are available from the various excitation sources,
and what energy levels are populated to a large enough extent to
actually absorb radiation.

Procedure

The literature data are generally published using one of the
following formats: (1) the limits of detection for each of the

3

lines investigated are listed (limits of detection generally being
defined as a signal-to-noise ratio of 2, which can be related to
the generally accepted statistical definition of limits of
detection)[51]; (2) the sensitivities for the lines under investigation are listed (the sensitivity is generally defined as the number
of parts per million of the element which causes 1% absorption, or
a 99% transmission); or, (3) the relative absorptions or relative
percent transmissions at each of the lines investigated are listed
for a particular concentration of analyte aspirated into solution.
In all three of these methods of data representation, the data are
either directly proportional or inversely proportional to the
relative sensitivity of the individual lines. In the atlas this
proportionality was set up such that the most intense line for an
element is set at a value of 10.   The value of 10 was chosen
rather than a value of 100 or 1000, thereby giving essentially one
significant figure. This was decided because of several factors.
First of all, there have been numerous improvements on hollow
cathode discharge lamps over the past several years; thus, the
data taken on the same instrument with an old hollow cathode lamp
may be quite different from those taken today with a new hollow
cathode lamp. Secondly, hollow cathode lamps produced by different
manufacturers often differ in relative intensity of emission lines,
thus causing minor changes in relative sensitivities of the lines.
Finally, different experimental conditions and setups have different sensitivities. It is felt, however, that one significant
figure is useful and should give a fairly good guide as to the
relative sensitivity of the various lines within an element.

     This atlas is not intended to be a literature survey of all
of the data of this type taken, and in fact many of the data have
been deliberately left out. Priorities were set up as to which
data to include. Because of the developments in atomic absorption
technology, the most recent articles have been given the highest
priority. Second, the thoroughness of the authors' investigations
has been taken into consideration. Finally, the reliability of
the data was considered.

Explanation of the Atlas

     The format of the atlas is relatively straightforward and
requires little explanation. All of the elements are listed

alphabetically according to name, and all of the elements from
helium through nobelium are included even though in several in-
stances there is no information for a particular element. The
authors believe it is often important to know that no information
is available. The wavelengths are given wherever possible to conform with those given in NBS Monograph 32.[42] For elements ex-
hibiting absorption both by the free atoms and by ions, the usual
designations I and II are given, respectively. No such designation
was given to elements that give rise only to transitions from free
atoms. In the cases where the wavelength data were not available
in the Monograph, [42] they were taken from the reference cited. The
relative absorption sensitivity is given on a scale from 10 down,
with 10 being the strongest absorbing line. Where possible, the
limit of detection or sensitivity cited in the reference is given
for the strongest absorbing line in the next column. The key for
the flame composition, or cell type, is given in Table 2. The key
for the type of hollow cathode, or other type of excitation source,
is given in Table 3. And finally, the listing under the reference
column refers to the reference number at the end of Section I.

Discussion

It should be strongly emphasized that the data presented in
this atlas do not represent every possible absorption line -- they
represent only those which have been experimentally observed using
the conditions described in the references, and while for some of
the elements the data are extremely complete, there are other elements for which the data are relatively incomplete or even non-
existent. The literature was covered through the year 1968 and
part of 1969 in this compilation. The elements for which rather
incomplete data exist are cesium, lead, manganese, niobium,
palladium, plutonium, praseodymium, rhenium, and ruthenium. It
is interesting that for eight elements -- chromium, cobalt,
gallium, platinum, thallium, vanadium, yttrium, and zirconium --
there is a difference of opinion in the literature as to which is
the strongest absorption line. Also for ten other elements --
argon, arsenic, gadolinium, gallium, indium, niobium, palladium,
rhenium, scandium, and sodium -- the literature reports at least
two lines with the same absorption sensitivity. There are only
two elements -- beryllium and mercury -- for which only one line
is given.

As Goleb[4] has shown, there is no reason to expect that the inert gases should not exhibit absorption. There is also no reason to expect that other gases such as hydrogen, oxygen, nitrogen, fluorine, chlorine, etc., could not be analyzed successfully by atomic absorption methods provided suitable sample cells and excitation sources could be arranged. Whereas atomic absorption has proven itself to be a useful tool for the rare earths, there is essentially no data to date for the transuranium elements, which should also be capable of analysis.

It is striking to observe that although there are admitted gaps in this atlas, data are presented for some seventy-odd elements of the 102 included.

TABLE 1

ATLAS OF ATOMIC ABSORPTION ANALYTICAL LINES*

Element	Symbol	Wavelength[1] (Å)	Relative[2] Absorption	Flame or[3] Cell	Lamp[4] Type	Limit of[5] Detection or Sensitivity (ppm)	Reference[6]
Actinium	Ac	No information					
Aluminum	Al	3092.71⎫ 3092.84⎭	10	A,D	ASL	(0.7)	1
		3961.53	8				
		3082.16	7				
		3944.03	5				
		2373.13⎫ 2073.36⎭	3				
		2367.06	2				
		2575.10⎫ 2575.41⎭	1				
		3092.71⎫ 3092.84⎭	10	B	PE	0.5	2
		3961.53	9				
		3082.16	7				
		3944.03	5				
Americium	Am	No information					
Antimony	Sb	2175.81	10	B	EDT	0.5	3
		2068.33	7				
		2311.47	5				
		2127.39	0.7				
Argon	Ar	8115.31	10	K	SG	NA (footnote 8)	4
		7635.11	10				
		8014.79	7				
		7514.65	6				
		8424.65	5				
		7383.98	5				
		7948.18	4				
		6965.43	3				

*Footnotes for this table can be found on page 31.

(Table 1 cont'd)

Argon (continued)		5048.81	3					
		8006.16	3					
		7067.22	3					
		7503.87	2					
		5054.18	2					
		5373.49	1					
		5221.27	1					
		3670.64	0.9					
		3675.22	0.9					
		3834.68	0.5					
		5659.13	0.4					
		8408.21	0.3					
		8103.69	0.2					
		4158.59	0.2					
		8521.44	0.2					
Arsenic	As	1936.96	10	B,C	EDT	0.5		3
		1890	10					
		1971.97	5					
Astatine	At	No information						
Barium	Ba	5535.48 I	10	A,J	ASL	(0.4)		5
		4554.03 II	NA					
		4934.09 II						
		5519.05 I						
Berkelium	Bk	No information						
Beryllium	Be	2348.61	10	A	PE	0.002		6
Bismuth	Bi	2230.61	10	NA	PE	0.2		7
		2228.25	5					
		3067.72	3					
		2061.70	1					
		1953.89 }						
		1959.48	0.8					
		2276.58	0.7					
		2110.26	0.4					
		2021.21	0.1					
Boron	B	2496.78	10	A,D	ASL	50		1
Bromine	Br	No information						

(Table 1 cont'd)

Cadmium	Cd	2288.02	10	<u>B</u>,E	ASL	(0.03)	8
		3261.06	0.02				
Calcium	Ca	4226.73 I	10	B	PE	~1	9
		2398.56 I	1				
		4454.78 I					
		3968.47 II	NA	J			5
		3933.67 II					
		5188.85 I					
		4425.44 I					
		4434.96 I					
		6161.29 I					
		6122.22 I					
		4226.73 I	10	B	PE	(0.5)	49
		3968.47 II	1				
		3933.67 II	1				
		2398.56 ND (Footnote 9)					
Californium	Cf	No information					
Carbon	C	No information					
Cerium	Ce[7]	No absorption lines to date with conventional equipment					1,10
Cesium	Cs	8521.10	10	B,<u>E</u>	ASL	(0.15)	11
Chlorine	Cl	No information					
Chromium	Cr	3578.69	10	F	HM	0.5	12
		3605.33	5				
		4254.35	5				
		4274.80	2				
		4289.72	2				
		3593.49	10	B	PE	(.08)	13
		3578.69	7				
		3605.33	6				
		4254.35	4				
		4274.80	3				
Cobalt	Co	2424.93	10	B	ASL	0.005	14
		2521.36	8				
		2411.62	6				
		2407.25	6				
		2432.21	4				
		2415.30	2				
		2435.83	2				

(Table 1 Cont'd)

Cobalt (continued)	2528.97	2				
	3044.00	1				
	2536.49	1				
	2439.05	1				
	3412.34					
	3412.63	0.8				
	3453.50	0.8				
	3526.85	0.7				
	3564.95	0.5				
	3405.12	0.4				
	2989.59	0.4				
	3513.48	0.4				
	2309.02	0.3				
	3431.58	0.3				
	3474.02	0.3				
	3506.32	0.3				
	2407.25	10	B	ASL	0.005	15
	2424.93	7				
	2521.36	4				
	2411.02	3				
	3526.85	0.4				
	3453.50	0.4				
	2295.23	0.3				
	3574.96					
	3575.36	0.3				
	3442.93	0.3				
	3502.28					
	3502.62	0.3				
	2987.16	0.2				
	3529.03	0.2				
	2274.49	0.1				
	2419.12	0.1				
Copper	Cu	3247.54	10	B,E	HM	1.1	16
		3273.96	5				
		2178.94	2				
		2165.09	1				
		2181.72	1				
		2225.70	0.6				
		2024.34	0.2				
		2492.15	0.1				
		2244.27	0.05				
		2441.64	0.02				
Curium	Cm	No information					

(Table 1 Cont'd)

Dysprosium	Dy	4211.72 I	10	A,G	PE	0.4	17,10
		4045.99 I	9				
		4186.78 I	8				
		4194.85 I	6				
		3531.70 II	5				
		3968.42 II	2				
		3645.41 II	2				
		3944.70 II	2				
		4167.99 I	2				
		4077.98 II	0.6				
		4000.48 II	0.3				

Einsteinium	Es	No information					

Erbium	Er	3372.76 II	10	A,G	PE	0.1	17,10
		4007.97 I	8				
		3862.82 I	6				
		4151.10 I	6				
		3264.79 II	5				
		3499.11 II	4				
		3892.69 II	3				
		3973.04⎫					
		3972.60⎭	2				
		4087.65 I	2				
		3937.02 I	2				
		3810.33 I	2				
		3312.42 II	2				
		3616.58 II	1				
		3692.64 II	0.9				
		3905.44 I	0.8				
		3944.41 I	0.7				
		4606.62 I	0.6				
		4409.35 I	0.4				
		4190.71 I	0.3				
		2985.50 I	0.2				
		4426.77 I	0.2				
		3558.02 I	0.05				

Europium	Eu	4594.03 I	10	A,G	PE	0.2	17,10
		4627.22 I	8				
		4661.88 I	7				
		4129.70 II	4				
		4205.05 II	3				
		3724.94 II	2				
		3210.57 I	0.8				
		3212.81 I	0.7				
		3111.43 I	0.7				
		3334.33 I	0.5				

(Table 1 Cont'd)

Fermium	Fm	No information					
Fluorine	F	No information					
Francium	Fr	No information					
Gadolinium	Gd	4078.70 I	10	A,G	PE	4	17,10
		3783.05 I	10				
		3684.13 I	10				
		4058.22 I	9				
		4053.64 I	8				
		3717.48 I	8				
		3713.57 I	6				
		4346.46 I ⎫					
		4346.62 I ⎭	6				
		4190.78 I	4				
		3679.21 I	4				
		4045.01 I	3				
		3362.23 II	2				
		3945.54 I	2				
		3266.73 I	2				
		3513.65 I	1				
		3358.62 II	1				
		3796.37 II	1				
		3768.39 II	1				
		3654.62 II	1				
		3423.90 I ⎫					
		3423.92 II ⎭	0.2				
Gallium	Ga	2874.24	10	G	HM	0.05	18
		2943.64	10				
		4172.06	6				
		4032.98	4				
		2500.70	1				
		2450.07	0.8				
		2943.64 ⎫					
		2944.18 ⎭	10	B	HM	1.3	19
		2874.24	10				
		4172.06	8				
		4032.98	6				
		2450.07	1				
		2500.17 ⎫					
		2500.70 ⎭	0.6				
		2719.65	0.6				

(Table 1 Cont'd)

Gallium (continued)	Ga	2943.64⎫ 2944.18⎭	10	A	HM	0.3	19
		2874.24	8				
		4172.06	7				
		2719.65	4				
		4032.98	4				
		2500.17⎫ 2500.70⎭	1				
		2450.07	0.7				
Germanium	Ge	2651.18⎫ 2651.58⎭	10	A	PE	1	6
		2592.54	5				
		2709.63	4				
		2754.59	3				
		2691.34	3				
		2041.69	2				
		2068.65	0.5				
		2497.96	0.5				
		2094.23	0.4				
		2651.18⎫ 2651.58⎭	10	D	ASL	(6.5)	1
		2592.54	4				
		2709.63	3				
		2754.59	3				
		2691.34	2				
		2651.18⎫ 2651.58⎭	10	A	HM	2	19
		2068.65	8				
		2041.69	8				
		2043.76	4				
		2198.70	0.8				
		2497.96	0.5				
		2065.20	<0.0004				
Gold	Au	2427.95	10	B	PE	0.05	20
		2647.95	6				
		3122.78	0.01				
		2748.26	0.009				
		6278.18	0.002				

(Table 1 Cont'd)

Hafnium	Hf	3072.88	10	A,D	ASL	25	1
		2866.37	8				
		2898.26	4				
		2964.88	3				
		3682.24	3				
		2950.68	2				
		3020.53	2				
		2904.41 ⎫					
		2904.75 ⎭	2				
		2940.77	2				
		3777.64	2				
Helium	He	3888.65	10	K	S-G	NA	4
		5875.62	8				
		6678.15	5				
		3187.74	3				
		3964.73	3				
		5015.68	3				
		4471.48	1				
		7281.35	0.8				
		4026.19	0.6				
		4921.93	0.6				
		3819.61	0.4				
Holmium	Ho	4103.84 I	10	A,G	PE	0.3	17,10
		4053.93 I	8				
		4163.03 I	6				
		4173.23 I	2				
		4040.81 I	2				
		4108.62 I	1				
		4127.16 I	0.9				
		3456.00 II	0.9				
		4101.09 I	0.8				
		4227.04 I	0.4				
		4136.22 I	0.3				
		4254.43 I	0.2				
		3955.73 I	0.2				
		5982.90 I	0.1				
		3796.75 I	<.06				
		3810.73 I	<.06				
		2518.73 II	<.06				
		3233.87 II	<.06				
		3398.98 II	<.06				
		3416.46 II	<.06				
		3484.84 II	<.06				
		3891.02 II	<.06				

(Table 1 Cont'd)

Hydrogen	H	No information					
Indium	In	3256.09	10	B	HM	.05	21
		3039.36	10				
		4104.76	4				
		4511.31	3				
		2560.15	0.8				
		2753.88	0.3				
		3256.09	10	A	HM	0.7	19
		3039.36	7				
		4511.31	2				
		4104.76	2				
		3258.56	0.7				
		2710.26	0.7				
		2560.15	0.6				
		2932.63	0.2				
		3256.09	10	B	HM	0.3	19
		3039.36	8				
		4511.31	3				
		4104.76	3				
		3258.56	1				
		2560.15	0.9				
		2932.63	0.4				
		2710.26	<.06				
Iodine	I	No information					
Iridium	Ir	2088.82	10	B	PE	2	22
		2639.42⎫	6				
		2639.71⎭					
		2664.79	5				
		2849.72	4				
		2372.77	4				
		2502.98	4				
		2092.63	3				
		2924.79	3				
		2475.12	2				
		2543.97	2				
		2881.16	1				
		3513.64	0.7				
		2661.98	0.6				
		3800.12	0.6				
Iron	Fe	2483.27	10	B	PE	(0.10)	13
		2522.85	5				
		2719.02	3				

(Table 1 Cont'd)

Iron (continued)		3020.49 ⎫ 3020.64 ⎭	3					
		2501.13	2					
		2166.77	2					
		3719.94	1					
		2966.90	1					
		3859.91	0.9					
		2983.57	0.8					
		3440.61	0.6					
		2936.90	0.6					
		2483.27	0.4					
		3824.44	0.1					
		3679.92	0.1					
Krypton	Kr	8059.50	10	K	SG	NA	4	
		8190.05	9					
		7601.54	8					
		7587.41	7					
		8112.90	7					
		8104.36	6					
		7694.54	5					
		7224.10	4					
		7854.82	2					
		7685.24	1					
		5870.91	1					
		4273.97	0.8					
		5570.28	0.8					
		7913.44	0.8					
		8298.11	0.4					
		8508.87	0.4					
		5562.22	0.3					
		8281.05	0.2					
Lanthanum	La	5501.34 I	10	A,G	PE	3	17,10	
		4187.32 I	7					
		4949.77 I	6					
		4086.72 II	4					
		3574.43 II	3					
		3649.53 I	2					
		3927.56 I	2					
		4037.21 I	2					
		4079.18 I	2					
		3613.08 I	2					
		4766.89 I	0.9					
		3898.60 I	0.6					
		5158.69 I	0.4					
		4662.51 II	<0.2					
Lead	Pb	2169.99	10	B,E	HM	0.3	23	
		2833.06	6					
		2614.18	0.06					

(Table 1 Cont'd)

Lead (continued)		2169.99	10	F	PE	0.45	24
		2833.06	7				
		4057.83	0.03				
Lithium	Li	6707.84	10	E	UH	0.03	23
		3232.61	0.06				
Lutetium	Lu	2615.42 II	10	A,G	PE	3	25
		3359.56 I	6				
		3312.11 I	4				
		3376.50 I	3				
		3567.84 I	3				
		3278.97 I					
		3281.74 I	3				
		3118.43 I	2				
		3396.82 I	0.9				
		2989.27 I	0.7				
		4518.57 I	0.5				
		3507.39 II	0.4				
		3081.47 I	0.3				
Magnesium	Mg	2852.13 I	10	B	W	(.01)	26
		2795.53 II	0.07				
		2802.70 II	0.04				
		5183.62 I	NA	J			5
		5172.70 I	NA				
		5528.46 I	NA				
		4703.02 I	NA				
		2852.13 I	10	B	PE	(0.8)	49
		2795.53 II	2				
		2802.70 II	ND				
Manganese	Mn	2794.82	10	B		(1.3)	27
		2798.27	8				
		2801.06	5				
		4030.76	0.8				
Mendelevium	Md	No information					
Mercury	Hg	2536.52	10	B,E	Hg vapor (10)		23

(Table 1 Cont'd)

Molybdenum	Mo	3132.59	10	B	HM	0.5	28
		3170.33	7				
		3798.25	6				
		3193.97	6				
		3864.11	5				
		3902.96	3				
		3158.16	3				
		3208.83	1				
		3112.12	0.4				
Neodymium	Nd[7]	4634.24	10	A,G	ASL	(35)	1,10
		4896.93	8				
		4719.02	5				
Neon	Ne	6402.25	10	K	SG	NA	4
		6143.06	6				
		6382.99	5				
		6334.42	4				
		6096.16	4				
		6266.49	3				
		7032.41	3				
		5944.83	3				
		6074.33	2				
		6929.47	2				
		6678.28	2				
		6163.59	2				
		5881.89	2				
		5852.48	2				
		6598.95	2				
		7245.16	1				
		6217.28	1				
		8377.61	0.9				
		6029.99	0.7				
		7438.90	0.6				
		8300.90	0.5				
		8136.41	0.5				
		7173.93	0.3				
		8418.43	0.2				
		7024.05	0.2				
Neptunium	Np	No information					

(Table 1 Cont'd)

Nickel	Ni	2320.03	10	B	HM	(0.13)	23
		2310.96	3				
		2345.54	0.6				
		3524.54	0.5				
		3414.76	0.4				
		3050.82	0.4				
		3002.49	0.2				
		3461.65	0.2				
		2320.03	10	NA	NA	(0.12)	13
		2310.96	7	(B)?	(PE)?		
		2345.54	3				
		2289.98	2				
		2337.49	0.6				
		3369.57	0.5				
		2347.52	0.4				
		3232.96	0.3				
		3391.05	0.3				
		3437.28	0.2				
		2476.87	0.04				
Niobium	Nb	4079.73 I	10	A	HB	5	6
		4058.94 I	10				
		4100.40 I⎫					
		4100.92 I⎬	8				
		4123.81 I	8				
		3726.24 I⎫					
		3727.23 I⎬	6				
		3790.15 I⎫					
		3791.21 I⎬	5				
		4168.13 I	5				
		3739.80 I	4				
		4163.66 I⎫					
		4164.66 I⎬	4				
		3798.12 I	3				
		4152.58 I	3				
		3787.06 I⎫					
		3787.48 I⎬	2				
		3759.55 I⎫					
		3760.64 I⎬	2				
		3802.92 I	2				
		4137.10 I⎫					
		4137.59 I⎬	1				
		2697.06 II	0.8				
		3741.78 I⎫					
		3742.39 I⎬	0.6				
		3163.40 II	0.3				
		3130.79 II	0.3				

(Table 1 Cont'd)

Niobium (continued)		3094.18 II	0.3				
		2927.81 II	<0.1				
		2950.88 II	<0.1				
Nitrogen	N	No information					
Nobelium	No	No information					
Osmium	Os	2909.06	10	A	PE	0.5	29
		3058.66	6				
		2637.13	6				
		3301.56	3				
		2714.64	2				
		2806.91	2				
		2644.11	2				
		4420.47	0.6				
		4260.85	0.3				
		2909.06	10	A	W	1	30
		3058.66	6				
		3018.04	4				
		3301.56	3				
		3267.94	2				
		3232.06	0.7				
		3262.29	0.4				
		3030.70	0.1				
Oxygen	O	No information					
Palladium	Pd	2447.91	10	H	HW	~2	31
		2476.42	6				
		3404.58	NA	H	HW	~1.2	32
		3242.70	NA				
		3634.70	NA				
		3609.55	NA				
		2476.42	NA				
		2447.91	10	B	PE	0.02	13
		2476.42	10				
		3404.58	3				
		2763.09	3				

(Table 1 Cont'd)

Element	Sym	Wavelength	Intensity	Col	Col	Col	Ref
Phosphorus	P	1774.95	10	N	HM	NA	13
				(Requires Vacuum Monochromation)			
Platinum	Pt	2853.11	10	F		10	12
		2659.45	10				
		2929.79	2				
		3064.71	0.3				
		2997.97	10	H	HM	5	32
		2646.89	10				
		2677.15	10	B	PE	0.1	13
		2174.67	7				
		3064.71	5				
		2628.03	4				
		2144.23	3				
		2830.30	3				
		2929.79	3				
		2733.96	2				
		2702.40	2				
		2705.89	2				
		2487.17	2				
		2646.89	2				
		2997.97	2				
		2067.50	2				
		2719.04	1				
		2677.15	1				
		2467.44	0.9				
		3042.64	0.6				
		2049.37	0.4				
		2450.97	0.4				
		2440.06	0.2				
		2165.17	0.07				
		3315.05	0.06				
Plutonium	Pu	4206.47	vvs[10]	M	NA	NA	33
		3878.52	vs				
		4208.24	vs				
		4735.40	vs				
Polonium	Po	No information					
Potassium	K	7664.91	10	E	ASL	(0.03)	11
		4044.14	0.06				

(Table 1 Cont'd)

Praseodymium Pr	4951.36	10	A,G	PE	10	17,10
	4914.03	7				
	5133.42	6				
	5045.53	3				
	4951.36	10	A	ASL	(72)	1
	5133.42	8				
	4736.69	5				
Promethium Pm	No information					
Protactinium Pa	No information					
Radium Ra	No information					
Radon Rn	No information					
Rhenium Re	3460.46	10	A	ASL	(12)	1
	3464.73	6				
	3451.88	4				
	3460.46	10	G	W	25	34
	3464.73	10				
	3451.88	5				
Rhodium Rh	3434.89	10	B	PE	0.03	35
	3692.36	6				
	3396.85	4				
	3502.52	3				
	3657.99	2				
	3700.91	1				
	3507.32	0.4				
	3434.89	10	A	PE	0.7	36
	3692.36	5				
	3396.85	4				
	3502.52	3				
	3507.32	3				
	3657.99	3				
	3700.91	1				
Rubidium Rb	7800.23	10	E	ASL	(0.1)	11
	4201.85	0.1				

(Table 1 Cont'd)

Ruthenium	Ru	3728.03	10	B	CF	(0.25)	37
		3498.94	2				
Samarium	Sm	4296.74	10	<u>A</u>,G	PE	5	<u>17</u>,10
		5200.59	6				
		4760.27	4				
		4728.42	4				
		5282.91	3				
		5117.16	3				
		4883.77⎫					
		4883.97⎭	3				
		4783.10	3				
		4581.58	2				
		5271.40	1				
		5403.70	0.8				
		5341.29	0.6				
		5320.60	0.6				
Scandium	Sc	3911.81	10	A	ASL	(0.8)	1
		3907.49	7				
		4023.69	7				
		4020.40	5				
		3269.91	3				
		3911.81	10	A	PE	0.2	17
		3907.49	10				
		4023.69	7				
		4020.40	5				
		4054.55	4				
		3269.91	3				
		4082.40	1				
		3273.63	0.8				
		3911.81	10	G	CA	(5)	34
		3907.49	10				
		4023.69	10				
		4020.40	10				
		3269.91	5				
		3273.63	5				
		4054.55	1				
		4082.40	1				
		3255.69	1				
		3996.61	1				
		3933.38	0.5				

(Table 1 Cont'd)

Element	Sym	Wavelength	Intensity	Ref	Method	Sens	No.
Selenium	Se	1960.26	10	B̲,C	LDT	0.1	3
		2039.85	2				
		2062.79	0.4				
		1960.26	10	H	EDT	(1)	38
		2039.85	5				
		2062.79	1				
		2074.79	0.4				
		2164	0.4				
Silicon	Si	2516.11	10	A	PE	0.6	38
		2506.90	4				
		2528.51	3				
		2514.32	3				
		2524.11	3				
		2216.67	3				
		2519.21	2				
		2210.88	2				
		2207.97	0.7				
		2516.11	10	D	ASL	10	1
		2514.32	8				
		2519.21	7				
		2506.90	4				
		2528.51	4				
		2524.11	3				
Silver	Ag	3280.68	10	E	ASL	(.05)	11
		3382.89	3				
Sodium	Na	5889.95	10	F	HM	1	12
		5895.92	10				
		3302.32	0.3				
		6160.76	NA	J	NA		5
		6154.23	NA				
		5149.09	NA				
		5153.64	NA				
		5688.22	NA				
		5682.66	NA				
Strontium	Sr	4607.33 I	10	E	ASL	(0.15)	11
		4077.71 II	0.6				
		4607.33 I	10	B	PE	(0.2)	49
		4077.71 II	5				
		4215.52 II	0.7				
		2569.47 I	0.5				

(Table 1 Cont'd)

Strontium (continued)	Sr	2931.83 I	0.2					
		2428.10 I	0.1					
		6892.59 I	ND					
Sulfur	S	No information						
Tantalum	Ta	2714.67 I	10	A,D	ASL	(30)	1	
		2607.84 II⎫						
		2608.20 I ⎬	5					
		2608.63 I ⎭						
		2775.88 I	5					
		2559.43 I	4					
		2647.47 I	3					
		2661.89 I⎫						
		2662.10 I⎬	2					
		2758.31 I	2					
Technetium	Tc	No information						
Tellurium	Te	2142.75	10	I	W	(0.4)	39	
		2259.04	1					
		2385.76	0.05					
		2142.75	10	B	PE	NA	40	
		2259.04	2					
		2383.25	0.5					
		2530.70	0.06					
		2769	0.06					
Terbium	Tb	4326.47	10	A,G	PE	2	17,10	
		4318.85	8					
		3901.35	6					
		4061.59	6					
		4338.45	5					
		4105.37	3					
Thallium	Tl	2767.87	10	E	ASL	(0.8)	11	
		3775.72	3					
		3775.72	10	B	PE	(1.2)	49	
		2767.87	5					
		2580.14	5					
		2379.69	3					
		2826.16	1					
		5350.46	1					
		2709.23⎫	0.4					
		2710.67⎭						

(Table 1 Cont'd)

Thallium (continued)	Tl	2918.32⎫ 2921.52⎭	0.4				
		2315.98	ND				
		2608.99	ND				
		2665.57	ND				
		3229.75	ND				
Thorium	Th	No absorption detected					1
Thulium	Tm	3717.92 I	10	A,G	PE	0.2	25
		4105.84 I	7				
		3744.07 I	6				
		4094.19 I	6				
		4187.62 I	5				
		4203.73 I	3				
		3883.13 I ⎫ 3883.44 II⎬ 3887.45 I ⎭	3				
		4359.93 I	1				
		2973.22 I ⎫ 2973.39 II⎭	0.8				
		3410.05 I	0.7				
		5307.12 I	0.5				
		4386.43 I	0.4				
		4733.34 I	0.3				
		3761.91 II	0.3				
		3996.62 I	0.3				
		3291.00 II	0.2				
		5675.85 I	0.2				
		3916.48 I	0.2				
		5631.40 I	0.1				
		3949.28 I	0.1				
		3416.59 I	0.06				
		5964.30 I	0.04				
Tin	Sn[7]	2246.05	10	B	EDT	0.5	3
		2354.84	7				
		2863.33	6				
		2429.49	3				
		2546.55	2				
		2839.99	--				
		2246.05	10	I	EDT	0.5	3
		2863.33	7				
		2354.84	5				
		2546.44	3				
		2429.49	1				

(Table 1 Cont'd)

Titanium	Ti	3653.50 I	10	A	HB	0.5	41
		3642.68 I	9				
		3199.92 I	8				
		3371.45 I	8				
		3752.86 I	6				
		3191.99 I	6				
		3741.06 I } 3741.64 II	6				
		3354.64 I	6				
		3186.45 I	5				
		2956.13 I	5				
		2646.64 I	4				
		3341.88 I } 3341.88 II	4				
		3998.64 I	4				
		2948.26 I	4				
		2644.26 I	4				
		2641.10 I	3				
		3989.76 I	3				
		2942.00 I } 2942.90 II	3				
		3635.20 I} 3635.46 I	3				
		3958.21 I	3				
		3729.82 I	3				
		2611.48 I	3				
		3981.76 I} 3982.48 I	3				
		3377.48 I} 3377.58 I	2				
		2605.15 I	2				
		3956.34 I	2				
		3948.67 I	2				
		3947.78 I	0.8				
		3653.50 I	10	D	ASL	(3)	1
		3642.68 I	10				
		3354.64 I	9				
		3199.99 I	9				
		3635.20 I} 3635.46 I	8				
Tungsten	W	2551.35	10	A	PE	3	17
		2681.41	8				
		2944.40	7				
		2946.98} 2947.38	5				
		4008.75	4				
		2831.38	4				

(Table 1 Cont'd)

Tungsten (continued)	2724.35	4					
	2996.45	4					
	2879.39	3					
	2656.54	3					
	4302.11	3					
	4074.36	2					
	4294.61	2					
	2606.39⎫						
	2606.90⎭	2					
	3617.52	1					
	2911.00	1					
	4269.39	0.2					
	2762.34	0.08					
	2452.00	0.08					
	3191.57	0.08					
	3768.45	0.08					
	2551.35	10	D		ASL	(17)	1
	2681.41	9					
	2724.35	6					
	4008.75	6					
	2946.98⎫						
	2947.38⎭	6					
	2831.38	5					
	2896.45						
	2656.54	4					
	2718.90	4					
	4074.36	4					
	2944.40	3					
Uranium U	3584.88 I	10	A		PE	30	17
	3566.60 I	7					
	3514.61 I	4					
	3943.82 I	4					
	3489.37 I	3					
	4153.97 I	3					
	3659.16 I	3					
	4042.76 I	3					
	3550.82 II	2					
	3500.07 I	2					
	3507.34 I	2					
	3890.36 II	2					
	3638.20 I	1					
	4050.05 II	0.8					
	4341.69 II	0.6					
	3670.07 II	0.4					
	4062.55 II	<0.4					
	4116.10 II	<0.4					
	2889.63 II	<0.4					
	3584.88 I	10	A		ASL	120	1
	3514.61 I	4					
	3489.37	4					
	3466.30	3					

(Table 1 Cont'd)

Vanadium	V	3185.40	10	A	HB	0.02	41
		3183.41⎫ 3183.98⎭	8				
		3060.46	4				
		3066.38	3				
		3056.33	3				
		4379.24	2				
		3703.58	2				
		4384.72	2				
		3855.37⎫ 3855.84⎭	2				
		3840.44⎫ 3840.75⎭	2				
		4389.97	2				
		4115.18	1				
		3053.65	1				
		3902.25	1				
		4111.78	1				
		3828.56	1				
		3704.70	1				
		2526.22	0.8				
		2923.62	0.8				
		2519.62	0.6				
		2574.02	0.5				
		2517.14	0.3				
		2530.18	0.3				
		2507.78	0.2				
		2511.65⎫ 2511.95⎭	0.2				
		3183.98	10	D	ASL	(2)	1
		3185.40	8				
		3183.41	8				
		3066.38	2				
		3703.58	2				
		3840.44⎫ 3840.84⎭	2				
		4379.24	1				
		3855.37⎫ 3855.84⎭	1				
		4384.72	0.8				
		4389.97	0.8				
Xenon	Xe	8231.63	10	K	SG	NA	4
		2280.12	9				
		8206.34	5				
		8409.19	5				
		6498.71	3				
		7386.00	2				

(Table 1 Cont'd)

Xenon (continued)		4671.23	2	K	SG	NA	4
		6198.26	2				
		7967.34	2				
		8346.82	1				
		8266.51	1				
		4624.28	0.8				
		4807.02	0.4				
		7887.39	0.3				
		4829.71	0.3				
Ytterbium	Yb	3987.98 I	10	**A**,G	ASL	(0.25)	1,10
		3464.36 I	3				
		2464.49 I	2				
		2672.65 II	0.2				
Yttrium	Y	4102.38	10	A	ASL	(5)	1
		4128.31	9				
		4077.38	9				
		4142.85	4				
		4077.38	10	G	CA	(50)	34
		4102.38	5				
		4128.31	5				
		4142.85	5				
Zinc	Zn	2138.56	10	E	ASL	(0.03)	11
		3075.90	0.002				
Zirconium	Zr	3547.68	10	A	PE	5	17
		3011.75	8				
		2985.39	8				
		3029.52	8				
		3623.86	7				
		3601.19	10	D	ASL	(18)	1
		3519.60	9				
		3011.75	6				
		3863.87	5				
		3547.68	4				
		3623.86	4				
		3029.52	3				
		2985.39	3				
		3890.32	3				
		3509.32	2				

Table 1

FOOTNOTES

1. Where possible the wavelengths conform to those given in NBS
    Monograph 32 (Reference 42). For elements exhibiting absorp-
    tion from both the free atoms and ions, the designation I and
    II is given respectively; in all other cases the wavelength
    refers to transitions arising from free atoms.

2. The relative absorption values refer to each specific element
    and cannot be used to compare absorption between different
    elements, unless reference is made to the limit of detection
    or sensitivity. These values are dependent on flame type and on
    conditions and should only be used as a guide. See the text
    for a more complete discussion of these values.

3. The key to the flame types and sample cells is given in Table
    2 (p. 33). In cases where two flame types or sample cells are
    indicated, the relative absorption values refer to the one that
    is underlined.

4. The key to the lamp type is given in Table 3 (p. 34).

5. Sensitivity values are in parentheses; limit of detections
    are not.

6. In cases where two references are given, the relative absorp-
    tion values refer to the one that is underlined.

7. F. W. Paul (Reference 43) observed 550 Ce lines, 880 Sn lines,
    and 300 Nd lines by means of a modified king furnace; however,
    he gave no indication as to the relative absorption of these
    lines.

8. NA means the information was not available in the reference.

9. ND means not detected.

10. VVS means very, very strong, VS means very strong. The author
    also gave many other lines of lesser intensity.

Table 2

KEY FOR FLAME CELL TYPE USED IN TABLE 1

A. Nitrous Oxide - Acetylene

B. Air - Acetylene

C. Entrained Air - Nitrogen - Hydrogen

D. Oxygen - Nitrogen - Acetylene

E. Air - Coal Gas

F. Oxygen - Hydrogen

G. Oxygen - Acetylene

H. Air - Propane

I. Air - Hydrogen

J. Furnace   (Reference 5)

K. Quartz Tube (Reference 4)

L. Furnace (Reference 43)

M. Furnace (Reference 44)

N. Sputtering Chamber   (Reference 45)

Table 3

LAMP TYPE

(Key for Excitation Source Type Used in Table 1)

ASL   Atomic Spectra Lamp, hollow cathode discharge tube (HCDT)
        (P & Y Ltd., Victoria, Australia)

PE   Perkin-Elmer   HCDT

SG   Schulet-Gollnow, water cooled HCDT (Reference 46)

HM   Home made HCDT (may or may not be described in reference
        cited)

W   Westinghouse HCDT

HB   High brightness HCDT   (Reference 48)

PEHB   Perkin-Elmer high brightness HCDT

HW   Hilger & Watts HCDT

UH   Hollow cathode discharge tube of unknown origin

CA   Continuous source, 150 W Xeron arc

CF   Continuous source, tungsten filament

EDT   Electrodeless discharge tube (Reference 47)

Section I

REFERENCES

1. M. D. Amos and J. B. Willis, Spectrochim. Acta, 22, 1325 (1966).

2. D. C. Manning, Atomic Absorption Newsletter, 3, 6 (1964).

3. R. M. Dagnall, K. C. Thompson, and T. S. West, Atomic Absorption Newsletter, 6, 117 (1967).

4. J. A. Goleb, Anal. Chem., 38, 1059 (1966).

5. G. D. Liveling and J. Dewar, "Collected Papers on Spectroscopy," Cambridge University Press, London, 1915, pp. 30-31.

6. D. C. Manning, Atomic Absorption Newsletter, 6, 35 (1967).

7. D. C. Manning, J. Vollmer, and F. Fernandez, Atomic Absorption Newsletter, 6, 17 (1967).

8. L. Wilson, Anal. Chim. Acta, 35, 123 (1966).

9. G. Patterson, Atomic Absorption Newsletter, 5, 117 (1966).

10. V. G. Mossotti and V. A. Fassel, Spectrochim. Acta, 20, 1117 (1964). (Many more lines are reported in this reference; however, only a rough indication of the relative absorbance is given.)

11. B. M. Gatehouse and J. B. Willis, Spectrochim. Acta, 17, 710 (1961).

12. J. W. Robinson, Anal. Chem., 33, 1067 (1961).

13. W. Slavin, "Atomic Absorption Spectroscopy," Interscience, New York, 1968.

14. W. W. Harrison, Anal. Chem., 37, 1168 (1965).

15. J. S. Cartwright and D. C. Manning, Atomic Absorption News-letter, 5, 114 (1966).

16. J. E. Allan, Spectrochim. Acta, 17, 459 (1961).

17. D. C. Manning, Atomic Absorption Newsletter, 5, 127 (1966).

18. C. E. Mulford, Atomic Absorption Newsletter, 5, 28 (1966).

19. R. E. Popham and W. G. Schrenk, Spectrochim. Acta, 24B, (1969).

20. D. C. Manning and J. Vollmer, Atomic Absorption Newsletter, 6, 38 (1967).

21. C. E. Mulford, Atomic Absorption Newsletter, 5, 28 (1966).

22. D. C. Manning and F. Fernandez, Atomic Absorption Newsletter, 6, 15 (1967).

23. W. T. Elwell and J. A. F. Gidley, "Atomic Spectrophotometry," Anchor Press, Ltd., 1966.

24. C. L. Chakrabarti, J. W. Robinson, and P. W. West, Anal. Chim. Acta, 34, 269 (1966).

25. F. Fernandez and D. C. Manning, Atomic Absorption Newsletter, 7, 57 (1968).

26. T. L. Chang, T. A. Gover, and W. W. Harrison, Anal. Chim. Acta, 34, 17 (1966).

27. J. E. Allan, Spectrochim. Acta, 15, 800 (1959).

28. D. J. David, Analyst, 86, 730 (1961).

29. F. Fernandez, Atomic Absorption Newsletter, 8, 90 (1969).

30. W. Osolinski and N. H. Knight, Appl. Spectr., 22, 532 (1968).

31. G. Erinc and R. J. Magee, Anal. Chim. Acta, 31, 197 (1964).

32. A. Strasheim and G. J. Wessels, Appl. Spectr., 17, 65 (1963).

33. L. Borey, Spectrochim. Acta, 10, 383 (1957).

34. V. A. Fassel and V. G. Mossotti, Anal. Chem., 35, 252 (1963).

35. P. Heneage, Atomic Absorption Newsletter, 5, 64 (1966).

36. M. G. Atwell and J. Y. Herbert, Appl. Spectr., 23, 480 (1969).

37. J. E. Allan, Spectrochim. Acta, 18, 259 (1962).

38. C. S. Raun and A. N. Hambly, Anal. Chim. Acta, 32, 346 (1965).

39. C. L. Charabarti, Anal. Chim. Acta, 39, 293 (1967).

40. J. Y. L. Wu, A. Droll, and P. F. Lott, Atomic Absorption Newsletter, 7, 90 (1968).

41. J. S. Cartwright, C. Sebens, and D. C. Manning, Atomic Absorption Newsletter, 5, 91 (1965).

42. W. F. Meggers, C. H. Corliss, and B. F. Scribner, "Tables of Spectral Line Intensities," N. B. S. Monograph 32, U. S. Government Printing Office, 1961.

43. F. W. Paul, Phys. Rev., 47, 799 (1935).

44. G. K. T. Conn and C. K. Wu, Trans. Faraday Soc., 34, 1483 (1938).

45. A. Walsh, "Colloq. Spectros. Intern. 10th," University of Maryland, 1962.

46. S. Tolansky, "High Resolution Spectroscopy," Methuen, London, 1947.

47. J. M. Mansfield, M. P. Bratzell, H. O. Norgordon, D. N. Knapp, K. E. Zacha, and J. D. Winefordner, Spectrochim. Acta, 23B, 389 (1968).

48. J. V. Sullivan and A. Walsh, Spectrochim. Acta, 21, 721 (1965).

49. Unpublished data obtained in the authors' laboratories.

50. M. Margoshes, Anal. Chem., 39, 1093 (1967).

51. M. L. Parsons, J. Chem. Ed., 46, 290 (1969).

SECTION II

ATOMIC EMISSION SPECTRAL LINES

Section II

ATLAS OF ATOMIC EMISSION SPECTRAL LINES

Introduction

This section was included in this book essentially for the sake of completeness. Dr. P. T. Gilbert Jr. has prepared such an atlas for most elements which emit in air-hydrogen, oxyhydrogen, and oxyacetylene flames. His compilation appears in the Handbook of Chemistry and Physics (Footnote 4, Table 4), and many of the data in this atlas are taken from that reference; however, the format has been rearranged to conform more closely to the format of Section I. The relative intensities are inclusive for each element in this section. Gilbert's data are essentially complete for oxyhydrogen and air-hydrogen flames; however, some of his data for oxyacetylene flame have been updated, with the appropriate reference given. Essentially all the update work was by Dr. Fassel's group (Footnote 5, Table 4). Mavrodineanu and Boiteux (Footnote 5, Table 4) have done extensive work with the air-acetylene flame. Unfortunately, they chose to use spectrographic plates rather than photoelectric detection, and their data will not necessarily correlate as well as those from the other workers.

In the last two to three years a new flame type has become more prominent in flame emission spectroscopy, the nitrous oxide-acetylene flame. This flame was not included in the atlas, however, because the data at this point are too sketchy.

Procedure

The procedure for this section is similar to that in Section I in that the most intensely emitting line of each element is given an arbitrary designation of 10. All the other lines are set proportionately lower according to their relative intensity. In the case of the air-acetylene flame for which data were taken from spectrographic plates, arbitrary numerical values were given to correspond to the designations in Mavrodineanu's book (see foot-

41

notes 5 and 8, Table 5). Again, it was felt that it was important
to point out if no information is available for a particular
element.

Finally, as Gilbert did not give indications as to the limit
of detection or sensitivity in his compilation and the book by
Mavrodineanu does not directly correlate to the general photoelectric detector systems, it was decided to pick arbitrarily from the
available literature the limits of detection for particular lines
and to intersperse them throughout the table as an indicator of
how sensitive a particular element and/or flame type is. It must
be emphasized that these limits of detection are not from the same
reference as the relative intensity data in most cases; therefore,
they should be taken only as indicators of the sensitivity of the
particular flame.

Explanation of the Atlas

Again, as in Section I, it is felt that this atlas is relatively straightforward and requires little explanation. However,
it is felt that two or three point should be made. The elements
are listed in alphabetical order according to name. The emitting
species is much more varied in flame emission spectroscopy, i.e.,
often the emitting species is a molecule of the atom rather than
the atom in question. Therefore, the symbol for the actual emitting species is placed under the symbol heading, and all lines that
are listed below the symbol are for the same species. For instance,
for aluminum there are some ten bands emitting from aluminum oxide
before the lines of atomic aluminum, which are given the usual
symbol of a Roman numeral I. This is further discussed in footnote
1, Table 4.

Again, where possible a single uniform reference is used for
the emitting species (see footnote 2, Table 4); however, in the
case of molecular emission this was often impossible.

Four flame types are considered: the oxyhydrogen, the air-
hydrogen, the oxyacetylene, and the air-acetylene flames. Under
each flame the relative intensity (RI) and the limit of detection
(LOD) are reported. In some cases the limits of detection were not
available for a particular flame type, and in those cases none is

given. There are also cases where no information was available
about a particular flame type, and this is indicated by the symbol
NI.

Discussion

     Infomation is presented for some 70 elements in this section.
It is interesting to note the differences between this section and
the section on atomic absorption. Only nine elements are listed
which are in complete agreement for all flames in both this section
and in the atomic absorption section for the most sensitive line.
These are the alkali metals plus germanium, mercury, ruthenium, and
silicon. Nineteen other elements list the same most sensitive line
for at least some of the flame types observed. However, different
lines were observed for certain flames in this group. Sixteen
different elements showed no agreement as to the most sensitive
line in flame emission as compared to atomic absorption, and
twenty-two elements show only band emission as the most sensitive
emission. Only in a very small number of cases is the most intense
emission the same as the most sensitive absorption in the flames so
far studied. Furthermore, in most cases where the sensitive lines
do correspond, they do so only in the hot flames such as the pre-
mixed-oxyacetylene flame. The large number of band emissions
observed as being analytically useful (thirty-six different band
emissions) indicates the importance of molecular species in the
flame.

     Although data are given for the same elements in these two
sections for the most part, there are several elements for which
no atomic absorption data have been obtained. These elements are
cerium, chlorine, iodine, promethium, ruthenium, and thorium.
Elements for which atomic absorption lines but no atomic emission
data are given are argon, helium, neon, plutonium, selenium, and
xenon. As in Section I, it should be strongly emphasized that the
relative intensities presented in the atlas are general guidelines
only; they should not be used in any absolute sense.

TABLE 4

ATLAS OF FLAME EMISSION LINES*

ELEMENT	SYMBOL[1]	WAVELENGTH[2]	OH[4]		AH[4]		OA[4,5]		AA[6,8]	
			RI	LOD[7]	RI	LOD[7]	RI	LOD[7]	RI	LOD[7]
Actinium	Ac	No information								
Aluminum	AlO	4842.3	10	2/A	}10		1.5		–	
		4866.4	10				1.25		–	
		4672.0	6.7		5.3		0.75		–	
		5079	6.7				0.75		–	
		5102	6.7		}6.7		0.90		–	
		5123	6.7				0.90		–	
		5143	6.7				0.85		–	
		4648.2	5.7		}5.3		0.65		–	
		4694.6	5.7				0.70		–	
		4715.6	5.7				0.65		–	
	Al I	3961.53	3.3		2.3		10	0.1/B	0.10	
		3944.03	2.7		2.0		5		0.01	
Americium	Am	No information								
Antimony	(SbO)	2574	NI		10		–		–	
	Sb I	2311.47			0.7	0.4/C	(10)	20*	2.0	
		2598.09			0.25		(10)		10	
		2528.52			0.10				10	

*Footnotes for this table can be found on page 70.

(Table 4 Cont'd)

Element	Species	Wavelength	OH[4] RI	OH[4] LOD[7]	AH[4] RI	AH[4] LOD[7]	OA[4,5] RI	OA[4,5] LOD[7]	AA[6] RI	AA[6] LOD[7]
Argon	Ar	No information								
Arsenic	(As) As I	5000	NI		10		(10)	100/A	-	
		2349.84			-		-		1.0	
		2288.12			-		-		0.30	
		2492.91			-		-		0.01	
	AsO	2504			-		-		-	
	As I	2745.00			-		-		0.10	
		2780.22			-		-		0.10	
		2860.44			-		-		1.0	
Astatine	At	No information								
Barium	Ba I	5535.48	10	0.25/A	8.5		6.7	0.05*	5.0	
	BaOH	4880	7.5		5.0		1.3		-	
		5130	7.5		7.5		2.0		-	
		8300	7.5		10		10	0.03/B	-	
		8730	7.5		4.0		10		0.01	
	Ba II	4934.09	6.3		4.0		3.3		-	
	BaOH	4970	6.3		4.0		1.1		-	
		5020	6.3		4.0		2.0		0.1	
		5240	6.3		4.0		1.3		0.01	
	BaO	5349.7	5.0		4.0		1.1		0.01	
		5492.7	5.0		4.0		1.0		-	
		5644.1	5.0		4.0		1.1		0.3	
		5701	5.0		4.0		1.1		0.3	
		5864.5	5.0		3.5		1.1		-	
		6039.6	5.0		3.5		4.7		0.10	
	BaOH	7450	2.5		2.5		-		-	
	Ba II	4554.03	2.0		0.4		3.3	0.03*	0.10	

(Table 4 Cont'd)

Element		Wavelength	OH RI	OH LOD	AH RI	AH LOD	OA RI	OA LOD	AA RI	AA LOD
Berkelium	Bk	No information								
Beryllium	BeO	4708.6	10	30/A	⎫		10		NI	
		4733	7.2		⎬ 10		8.0			
		4755	5.7		⎭		4.4			
		5054.4	3.6		⎫		4.0			
		5075.7	3.6		⎬ 5		4.0			
		5095.1	3.6		⎭		4.0			
	Be I	2348.61	–		–		10	1.0*	–	
Bismuth	Bi I	4722.55	10	20/A	10		10	40*	–	
	BiO	5564	4.0		10		2.0		–	
	(BiH)	4424	2.0		8.0		0.8		–	
		4394	1.6		8.0		0.6		7.0	
	Bi I	3067.72	0.7		4.0		2.0		–	
		2230.61	–		–		–			
		2276.58	–		–					
Boron	BO₂	5476.0	10	0.2/A	10		10	30*	5.0	
		5180.7	8.3		7.3		4.0		2.0	
		4941.3	5.0		4.5		3.3		0.3	
		5790.7	5.0		6.4		6.7		2.0	
		4719.5	3.3		2.3		1.7		0.3	
		6030	1.7		1.8		4.7		–	
		4530	1.2		0.9		0.8		–	
		6202.2	1.2		1.3		2.0		0.3	
	B I	2496.78	–		–		10		–	
Bromine	Br	No information								

(Table 4 Cont'd)

Element	Species	Wavelength	OH RI	OH LOD	AH RI	AH LOD	OA RI	OA LOD	AA RI	AA LOD
Cadmium	Cd I	3261.06	10	0.5/A	10	0.3/C	10	30*	7.0	
		2288.02	5		0.1		10	6.0*	0.1	
Calcium	CaOH	6220	10	0.004/A	10		10		10	
		5540.	6.8		10		3.4		10	
	Ca I	4226.73	4.0		5.0		5.0	0.005*	10	
	CaOH	6038	2.8		2.0		5.0		1.0	
		6453	1.2		1.4		1.4		2.0	
		5720	0.40		0.5		–		0.3	
	Ca II	3933.67	0.01		–		0.60	0.005*	–	
		3968.47	0.008		–		0.34		–	
	CaO	8720	VW		–		VW		–	
		8240	VW				VW			
Californium	Cf	No information								
Carbon	C	No information								
Cerium	CeO	5500-6000	10	1/B	10	0.15/A	10	10*	10	
		4940	7		5.7		10		–	
		4684	5		3.6		7.1		–	
		4810	5		4.3		7.1		–	
Cesium	Cs I	8521.10	10	0.1/A	10		10	0.008*	10	
		8943.50	3.0		3.0		3.0	8.0*	10	
		4555.36	0.25		0.20		<.01		1.0	
		4593.18	0.07		0.05		<.01		0.1	

(Table 4 Cont'd)

			OH		AH		OA		AA	
	Cl	4354	RI	LOD	RI	LOD	RI	LOD	RI	LOD
Chlorine			10	700/D	-		-		-	
Chromium	Cr I	4254.35	10	0.1/A	4.8		10		10	
		3578.69	8.0		1.6		10		7.0	
		4274.80	8.0		4.4		8.5	0.01/B	10	
		3593.49	7.0		1.4		8.5	0.1*	7.0	
		4289.72	7.0		4.0		6.0		10	
		5206.04	7.0		6.8		5.0		0.3	
		3605.33	5.5		1.2		7.5		7.0	
	CrO	5852.1	2.5		10		8.5	0.1*	0.3	
		6051.6	2.5		10		10		1.0	
		5356	2.0		8.0		5.0		-	
		5417	2.0		8.0		5.0		0.01	
		5564.1	2.0		8.0		5.0			
		5623	2.0		8.0		5.0		0.3	
		5794.4	2.0		10		7.5		1.0	
		6394.3	1.0		-		8.5			
		6850	1.0		-		7.5		-	
Cobalt	Co I	3526.85	10	0.5/A	7.8		10	1.0*	7.0	
		3412.34	7.1		8.9		4.3		7.0	
		3453.50	7.1		10		10		7.0	
		3405.12	5.7		8.9		7.1		7.0	
		3873.12	5.7		6.7		5.7		7.0	
		3502.28	4.9		6.7		4.3		10	
		5635	4.3				-		-	
(CoO)	Co I	3513.48	4.3		4.4		4.3		5.0	
		3575.36	4.3		4.4		3.6		5.0	
		3449.44	3.4		5.6		4.3		2.0	

(Table 4 Cont'd)

			OH		AH		OA		AA	
			RI	LOD	RI	LOD	RI	LOD	RI	LOD
Cobalt Continued	Co I	3465.80	3.4		4.4		4.3		7.0	
		3845.47	3.1		4.9		2.9		2.0	
		4121.32	3.1		5.6		2.9		2.0	
		3443.64	2.9		4.4		3.6		2.0	
		3594.87	2.9		3.8		2.9		1.0	
		3894.08	2.9		4.9		2.9		7.0	
		3506.32	2.6		6.7		2.9		2.0	
		3995.31	2.6		5.6		2.9		2.0	
		3569.38	2.3		4.4		2.4		5.0	
		3474.02	2.0		3.8		2.9		2.0	
		3433.04	1.3		2.2		2.1			
Copper	Cu I	3247.54	10	0.01/A	6.0		10	0.1*	10	
		3273.96	10		4.0		10	0.2*	10	
	CuOH	5370	1.7		10		0.60		-	
		5240	1.2		7.0		0.50		0.01	
	Cu I	5105.54	0.80		-		0.70		0.01	
	CuOH	5050	0.70		5.0		0.25		-	
	CuO	6050	0.50		-		0.30		-	
	CuH	4279.6	0.30		-		0.07		0.3	
Curium	Cm	No information								
Dysprosium	DyO	5729	10	0.08/B	NI		0.3	**	7.0	
		5833.8	10				0.3	**	10	
		5263	5.8				0.1	**	5.0	
		5398.2	5.0				-		2.0	
		5492.5	5.0				-		1.0	

(Table 4 Cont'd)

Element	Species	Wavelength	OH RI	OH LOD	AH RI	AH LOD	OA RI	OA LOD	AA RI	AA LOD
Dysprosium Continued	DyO	5204.4	3.0		NI		–			.01
		5160	2.4				–		–	
		6057	1.3				0.01	**	5.0	
		6091	1.3				–		–	
		4574.2	0.67				0.01	**	3.0	
	Dy I	4211.72	–				10	0.1*	–	
		4186.78	–				3.3	0.3*	–	
Einsteinium	Es	No information								
Erbium	ErO	5520	10	0.2/B	NI		0.01	**	.7	
		5459.3	8.0				0.01	**	.3	
		5600	7.0				0.01	**	.1	
		5041.3	6.0				0.01	**	0.1	
		5660	5.0				0.01	**	.3	
		5150.9	3.4				0.01	**	.01	
	Er I	4007.97	–				10	0.3*	–	
		4151.10	–				3.3	1.0*	–	
		5826.79	–				3.3	1.0*	–	
Europium	EuOH	5980.	10	0.1/B	NI		(10)		7.0	
	Eu I	6018.15	10				(10)		0.1	
	(EuOH)	6230	7.0				(10)		10	
		7020	5.0				(10)		10	
	Eu I	4594.03	2.5				10	0.003*	10	
	(EuOH)	6470	2.5				(10)		10	
	Eu I	4627.22	2.0				10		10	
	(EuOH)	6840	2.0				(10)		10	
	Eu I	4661.88	1.7				10		10	

(Table 4 Cont'd)

Element	Symbol	Wavelength (RI)	OH RI	OH LOD	AH RI	AH LOD	OA RI	OA LOD	AA RI	AA LOD
Fermium	Fm	No information								
Fluorine	F	No information								
Francium	Fr	No information								
Gadolinium	GdO	5910.8	10	0.12/B	(10)		(10)		3.0	
		5987.9	10		(10)		(10)		3.0	
		6120	8.8		(10)		(10)		7.0	
		6221.0	8.8		(5)		(5)		10	
		5807.5	6.3		(3)		-		2.0	
		5698.3	4.4		-		-		0.1	
		4615.8	2.1		-		-		7.0	
		4633.7	1.9		-		-		5.0	
		5450.7	1.9		-		-		-	
		4892.0	1.5		-		-		0.1	
		4909.5	1.5		-		-		0.01	
		5405.1	1.5		-		-		-	
		4928.0	0.56		-		-		-	
	Gd I	4519.66	-		10		10	2.0*	-	
		4346.46 }	-		5.0		5.0 }	4.0*	-	
		4346.62			-				-	
	Gallium Ga I	4172.06	10	0.05/A	10		10	0.07*	10	
		4032.98	5.0		5.0		5.0		10	
		2944.18	<.01		-		(0.1)		2.0	

(Table 4 Cont'd)

Element		Wavelength	OH RI	OH LOD	AH RI	AH LOD	OA RI	OA LOD	AA RI	AA LOD
Germanium	Ge I	2651.58	10	5/A	10	5/c	10	0.6*	-	
		2592.54	-		-		3.7		-	
		2709.63	-		-		2.5		-	
		2754.59	-				0.2			
Gold	Au I	2675.95	10	5/A	10	3.3	10	4*	0.3	
		2427.95	5.0		3.3		6.5		0.3	
Hafnium	Hf I	3682.24	NI		NI		10	75*	NI	
Helium	He	No information								
Holmium	HoO	5659	10	.08/B	NI		0.1	**	7.0	
		5157.0	4.2				0.01	**	1.0	
		5270.2	4.2				0.01	**	1.0	
		5319.6	4.2				0.01	**	1.0	
		5104.8	2.9				0.01	**	0.1	
		5849.4	2.9				0.01		1.0	
	Ho I	4103.84	-				10	0.1*	-	
Hydrogen	H	No information								

(Table 4 Cont'd)

Element	Species	Wavelength	OH RI	OH LOD	AH RI	AH LOD	OA RI	OA LOD	AA RI	AA LOD
Indium	In I	4511.31	10	0.03/A	10		10	0.03*	10	
		4104.76	5.7		6.0		7.1		10	
		3256.09	0.30		0.60		0.36		3.0	
	InO	3039.36	<.01		0.30		0.43		3.0	
		4283	<.01		-		-		-	
Iodine	IO	4694	(10)	10/A	NI		NI		NI	
		4845	(10)							
		4964	(10)							
		5131	(10)							
		5209	(10)							
		5308	(10)							
		5533	(10)							
		5730	(10)							
Iridium	Ir	3800.12	NI	0.2/A	NI		10	100*	-	
Iron	Fe I	3719.94	10		4.6		10	0.7*	7.0	
	FeO	5646.6	10		10		6.7		0.01	
		5789.8	10		} 10	0.06/B	6.7		0.01	
		5819.2	10				7.3		0.1	
		5868.1	8.75		8.7		6.7		1.0	
		5614.0	8.75		4.1		6.0		0.01	
	Fe I	3859.91	7.5		4.6		7.3		7.0	
		3737.13	6.25		4.1		5.3		7.0	
	(Fe I)	3747	5.0		5.9		4.0		5.7	
	FeO	5531.4	5.0		5.9		4.0		-	
		6097.3	5.0		5.9		8.0		0.1	
		6180.5					6.7		0.3	

(Table 4 Cont'd)

Element	Compound	Wavelength	OH		AH		OA		AA	
			RI	LOD	RI	LOD	RI	LOD	RI	LOD
Iron Continued	FeO	6218.9	5.0		5.9		6.7		0.1	
		5269.54	4.25		4.1		2.3		-	
		5328.05	3.75		4.1		2.3		-	
		3886.28	3.0		3.5		1.7		7.0	
		3825.88	2.75		3.5		1.1		7.0	
		3930.30	2.0		2.9		1.0		3.0	
		3878.58	1.75		3.5		0.8		2.0	
		3440.61	1.5		2.0		1.3		7.0	
		3581.20	1.5		2.9		0.8		2.0	
		3899.71	1.25		2.9		0.6		2.0	
		4383.55	1.25		2.9		0.6		0.01	
		3020.64	0.1		0.2		1.0		7.0	
Krypton	Kr	No information								
Lanthanum	LeO	5602.4	10	0.005/A	} 5.7		0.17		2.0	
		5628.6	7.2				0.17		0.3	
		4379.7	3.6		1.7		0.17		7.0	
		4423.2	3.6		1.7		0.17		7.0	
		5406	2.9		1.0		0.07		-	
		5431	2.9		1.0		0.07		-	
		7430	2.9		10		10		10	
		7920	2.4		8.3		10		10	
		5457	2.1		1.0		0.06		-	
		5382.5	1.4		1.0		0.05		-	
		5866	1.4		} 1.0		0.12		-	
		5896.7	1.4				0.12		0.01	
		5921	-		-		-		-	
		4400.0	-		-		-		7.0	
		7691.1							10	

(Table 4 Cont'd)

			OH		AH		OA		AA	
			RI	LOD	RI	LOD	RI	LOD	RI	LOD
Lanthanum Continued	LaO	7725.5	-		-		-		10	
		8159.0	-		-		-		10	
		8233.1	-		-		-		10	
		8270.7	-		-		-		10	
	La I	5791.34	-		-		10	1.0*	-	
Lead	Pb I	3683.48	10	1/A	10		6.7	3.0*	7.0	
		4057.83	10		10		10		7.0	
		3639.58	5.0		5.0		3.3		2.0	
		2833.06	0.4		-		-		2.0	
Lithium	Li I	6707.84	10	0.0002/A	10		10	.000003*	10	
		6103.64	<.01		0.03		0.01	.001*	0.3	
		4602.86	-		<.01		<.01		0.01	
Lutetium	LuO	5170.3	10	0.2/B	NI		0.3	**	-	
		4661.7	6.0				7.0	**	0.1	
		6000	4.0				-	**	0.1	
		6750	2.0				-		2.0	
	Lu I	3312.11	-		-		10	0.2*	-	
		4518.57	-		-		4.0	0.5*	-	
Magnesium	Mg I	2852.13	10	0.1/B	2.0	.02/A	10	0.2*	7.0	
	MgOH	3702	10		10		0.7		7.0	
		3810-3830	8.0		10		0.2		7.0	
		3877	5.0		4.0		0.1		2.0	
		3912	5.0		4.0		0.1		2.0	

(Table 4 Cont'd)

Element	Species	Wavelength	OH RI	OH LOD	AH RI	AH LOD	OA RI	OA LOD	AA RI	AA LOD
Magnesium Continued	MgOH	3624	2.5		2.0		0.07		0.3	0.3
	MgO	5007	1.0		-		0.1		-	0.1
	Mg I	5172.70	}0.5		-		0.04		0.1	
	Mg I	5183.62			-		-		0.3	0.3
Manganese	Mn I	4030.76, 4032.07, 4034.49	10							
	MnO	4032	10	0.01/A	10		10	0.1*	10	
		5586.5	0.8		2.4		0.34		7.0	
		5609.3	0.6		2.4		0.34		5.0	
		5389.5	0.5		1.0		0.16		2.0	
		5880.3	0.5		1.4		0.16		7.0	
		5359.4	0.45		1.0		0.16		1.0	
		5860	0.4		1.4		0.16		2.0	
		5423.7	0.4		1.0		0.14		1.0	
		5192.5	0.25		1.0		0.08		0.1	
		5228.4	0.25		1.0		0.07		0.1	
		6154.7	0.25		0.6		0.1		0.1	
		5158.0	0.2		1.0		0.06		0.01	
	Mn I	2794.82	0.15		-		0.03		2.0	
		2798.27	0.11		-		0.03		2.0	
	MnOH	3630-4100	0.1		1.4		-		-	
	Mn I	2801.06	0.08		-		0.03		2.0	
Mendelevium	Md		No Information							
Mercury	Hg I	2536.52	10	6/A	10	4/C	10	40*	7.0	
Molybdenum	(MoO$_2$)	5500-6000	1.0	1/A	1.0	0.4/B	(10)	0.03*	-	
	Mo I	3798.25	0.8		1.0		(.16)		10	
		3864.11	0.8		1.2		(.10)		7.0	
		3902.96	0.8		1.2		(.08)		7.0	

(Table 4 Cont'd)

Element	Species	Wavelength	OH RI	OH LOD	AH RI	AH LOD	OA RI	OA LOD	AA RI	AA LOD
Neodymium	NdO	5990	10	2/B	NI		1.0	1/A	10	0.01
		6600	10				10		10	
		6910	10				5.0		10	
		7020	10				10		10	
		7120	5.0				-		-	
		5314	3.0				-		-	
		4620	2.5				0.7		0.1	
		6220	2.5				2.0		2.0	
		6360	2.5				3.0		7.0	
		6430	-				-		10	
		6623.9	-				-		10	
		6942.6	-				-		-	
	Nd I	4883.81	-				10	1.0*	-	
		4924.53	-				10	1.0*	-	
Neon	Ne	No information								
Neptunium	Np	No information								
Nickel	Ni I	3414.76	10	0.3/A	10	0.12/B	6.7	0.6*	7.0	
		3524.54	10		10		10		7.0	
		3515.05	4.4		5.6		6.0		7.0	
		3619.39	4.4		5.0		3.3		7.0	
		3461.65	4.0		5.6		4.7		7.0	
		3392.99	3.4		3.1		3.3		7.0	
		3492.96	3.4		3.1		4.7		7.0	
	NiO	5200-6000	3.4		-		-		-	
		5174	3.0		-		-		-	

(Table 4 Cont'd)

		OH		AH		OA		AA	
		RI	LOD	RI	LOD	RI	LOD	RI	LOD
Nickel Continued	Ni I 3446.26	2.4		2.5		3.3		7.0	
	Ni0 5024	2.4		-		-		-	
	Ni I 3433.56	2.0		1.9		2.0		7.0	
	3566.37	2.0		2.5		2.0		2.0	
	3369.57	1.8		2.1		2.0		7.0	
	3380.57	1.4		1.5		1.6		2.0	
	3510.34	1.4		5.6		3.3		2.0	
	3610.46	1.4		1.9		1.6		0.3	
	3472.54	0.8		0.9		1.6		2.0	
	3858.30	0.8		1.3		1.2		2.0	
	3002.49	0.6		0.5		0.6		2.0	
Niobium	(NbO) 4500	10		10	2/A	5.7		-	
	5500	10		10		10		-	
	Nb I 4058.94	-		-		10	1.0*	-	
Nitrogen	No information								
Nobelium	No information								
Osmium	Os I 2909.06	NI		NI		NI		2.0	
	3018.04							1.0	
	3058.66							1.0	
	3301.56							1.0	
	4420.47					10	10*	-	

(Table 4 Cont'd)

Element	Symbol		OH		AH		OA		AA	
			RI	LOD	RI	LOD	RI	LOD	RI	LOD
Oxygen	O	No information								
Palladium	Pd I	3634.70	10	0.2/A	10		10	1.0*	10	
		3404.58	8.8		8.2		10		10	
		3609.55	5.6		5.4		7.0		10	
		3516.94	2.5		1.4		2.0		7.0	
		3421.24	1.3		2.0		2.0		7.0	
		3460.77	1.3		1.0		2.0		5.0	
Phosphorus	P O	5200	NI		10	1/c	NI		NI	
		2464			0.2					
		2375			0.1					
		2383			0.1					
		2478			0.1					
		2396			0.07					
		2529			0.07					
		2540			0.05					
Platinum	Pt I	3064.71	10	10/A	10		10	40*	7.0	
		2659.45	8.0		4.9		7.1		7.0	
Plutonium	Pu	No information								
Polonium	Po	No information								

(Table 4 Cont'd)

Element	Spectrum	Wavelength	OH RI	OH LOD	AH RI	AH LOD	OA RI	OA LOD	AA RI	AA LOD
Potassium	K I	7664.91	}10	0.0003/B	}10	0.03	10	0.003*	10	
		7698.98					6.7	0.02*	10	
		4044.14	<.01				<.01		2.0	
Praseodymium	PrO	5763	10		NI		-		-	
		5691	8.0				-		-	
		5610	7.3				-		-	
		5380	6.7				-		0.01	
		6030	6.7	10/A			1.0		7.0	
		6950	(6.7)				2.0	**	10	
		7095	(6.7)				2.0	**	7.0	
		7321	}(6.7)				0.01	**	7.0	
		7576						**	2.0	
		5157	4.7				-		0.01	
		6298	3.3				-		0.1	
		6363.1	3.3				-		0.3	
		6481.2	3.3				0.3	**	0.3	
		8050	(3.3)				0.3	**	5.0	
		8494.0	(1.3)						7.0	
	Pr I	4939.74	-				10	2.0*	-	
Promethium	(PmO)	(6400)	(10)		NI		(10)		NI	
		(6800)					(10)			
Protactinium	Pa	No information								
Radium	Ra I	4825.9	(10)		NI		NI		NI	
	(RaOH)	6270	(10)							

(Table 4 Cont'd)

Element	Species		OH RI	OH LOD	AH RI	AH LOD	OA RI	OA LOD	AA RI	AA LOD
Radium Continued	(RaOH)	6650	(10)							
	Ra II	3814.4	(6.0)							
		4682.3	(2.0)							
	(RaOH)	6020	(2.0)							
Radon	Rn		No information							
Rhenium	Re I	4889.14	(10)	100/A	NI		NI		NI	
		5275.56	(10)							
		3460.46	–		–		10	1.0*	–	
Rhodium	Rh I	3692.36	10	0.7/A	1.75		10	0.3*	10	
		3434.89	4.3		0.62		6.0		10	
	RhO	5425			10		0.67		–	
	Rh I	3657.99	4.3		1.50		4.0		10	
		3700.91	3.2		1.50		3.0		7.0	
		3396.85	3.2		1.0		3.3		7.0	
		3502.52	2.3		0.89		3.3		7.0	
		4374.80	2.3		3.75		1.1		2.0	
		3528.02	2.0		0.75		2.7		7.0	
		3596.19	1.4		1.25		1.4		7.0	
		4211.14	1.4		2.75		0.7		2.0	
		3583.10	1.1		1.13		1.1		2.0	
		3856.52	1.1		1.75		1.0		5.0	
		3323.09	0.7		0.50		1.1		5.0	

(Table 4 Cont'd)

			OH		AH		OA		AA	
		Wavelength	RI	LOD	RI	LOD	RI	LOD	RI	LOD
Rubidium	Rb I	7800.23	10	0.003/B	10		10	0.002*	10	
		7947.60	7.2		7.2		8.5		10	
		4201.85	0.1		0.06		0.01		0.3	
		4215.56	0.04		0.02		<.01		1.0	
Ruthenium	Ru I	5728.03	10	0.5/A	NI		10	0.3*	10	
		3799.35	5.0				8.3		10	
		3498.94	2.5				1.0		10	
		3428.31	1.0				0.5		7.0	
		3661.35	1.0				0.67		7.0	
		3436.74	0.5				0.33		7.0	
		3593.02	0.5				0.33		5.0	
		5699.05	0.25				0.33		-	
Samarium	SmO	6140	10	0.5/B	NI		6.67		2.0	
		5950	8.75				1.33		0.1	
		6030	8.75				1.67		0.3	
		5870	7.50				1.00		0.01	
		6240	7.50				8.33		7.0	
		5820	5.00				0.67		0.01	
		6380	5.00				6.67		7.0	
		6420.9	5.00				6.67		10	
		6520	5.00				10		-	
		4720	2.50				-			
		6680	1.25				1.67		7.0	
		6830	0.75				1.00		7.0	
	Sm I	4883.77	-				10	0.6*	-	
		5175.42	-				7.5	0.8*	-	

(Table 4 Cont'd)

			OH		AH		OA		AA	
			RI	LOD	RI	LOD	RI	LOD	RI	LOD
Scandium	ScO	6073	10	0.04/B	NI		10		7.0	
		6110	8.0				8.3		7.0	
		6017	} 6.8				} 6.7		} 7.0	
		6036								
		6154	4.4				3.3		7.0	
		6190	2.0				1.7		7.0	
		5812	1.6				1.7		2.0	
		5849	1.6				1.0		2.0	
		5773	1.4				1.0		1.0	
		5887	1.4				1.0		2.0	
		5928	1.2				1.0		0.3	
		6230	1.2				-		7.0	
		4858.1	0.8				0.7		1.0	
		5736.8	0.7				-		0.1	
		4893.3	0.6				-		0.1	
		4707.3	0.5				-		0.01	
		4673.1	0.4				-		0.1	
		6500	0.4				0.3		2.0	
		4742.4	0.3				-		-	
		5097	0.3				-		-	
		5134	0.3				-		-	
		5171	0.3				-		-	
	Sc I	3911.81	-				10	0.07*	-	
		4020.40	-				7.0	0.10*	-	
		3907.49	-				3.5	0.20*	-	
Selenium	Se	No information								
Silicon	Si	2516.11	NI		NI		10	5	NI	

(Table 4 Cont'd)

Element	Species	Wavelength	OH RI	OH LOD	AH RI	AH LOD	OA RI	OA LOD	AA RI	AA LOD
Silver	Ag I	3382.89	10		10	0.04/B	10	0.3*	10	
		3280.68	5.9	0.06/A	4.0		10		10	
Sodium	Na I	5889.95–5895.92	10		10		10		10	
		5688.22	<0.01		–	0.0001/A	<0.01	0.0001*	–	
		3302.99	<0.01		<0.01		<0.01		0.3	
		8194.81	<0.01		–		<0.01		7.0	
Strontium	Sr I	4607.33	10	0.01/A	1.0		10	0.004*	10	
	SrOH	6050	10		10		5.0		10	
		6820	10		0.5		1.5		10	
		6660	7.0		1.0		2.0		–	
	Sr₂O₂	6590	2.0		1.0		1.0		0.3	
		6450	1.0		0.5		0.5		0.3	
		5950	0.9		1.0		0.5		0.3	
		5970	0.8		1.0		0.5		0.3	
	SrOH	7070	0.7		–		–		0.3	
	Sr II	4077.71	0.3		–		0.9		0.3	
		4215.52	0.2		–		0.6		0.1	
Sulfur	S	No information								
Tantalum	Ta	4812.75	NI		NI		10	18*	NI	
Technetium	Tc	No information								

(Table 4 Cont'd)

Element	Species	Wavelength	OH RI	OH LOD	AH RI	AH LOD	OA RI	OA LOD	AA RI	AA LOD
Tellurium	TeO	3714	10		10		8.8		NI	
		3827	8.3		10		10			
		3884	8.3		10		10			
		3954	8.3		10		10			
		4131	7.3		10		10			
		4007	7.3		10		10			
		4075	7.3		8.6		10			
		4205	7.3		10		10			
		4268	7.3		8.6		10			
		4343	7.3		8.6		8.8			
		4487	6.8		10		8.8			
		3607	6.8		10		8.8			
		3662	6.8		7.2		10			
		3773	5.7		7.2		8.8			
		4640	5.0							
		3561								
	Te I	2385.25	} -		} 0.03		} 0.0	200*		
		2385.76						600*		
Terbium	TbO	5920.8	10	0.12/B	NI		(10)		10	
		5980	8.75				(10)		10	
		5730	6.25				-		2.0	
		5340	5.0				-		0.1	
		5639	5.0				-		0.1	
		6079.6	5.0				(10)		7.0	
		5440	3.75				-		-	
		6350	2.15				(5)		0.3	
		4610	1.00				-		0.3	
	Tb I	4326.47	-				10	1.0*	-	
		3901.35	-				2.5	4.0*	-	

(Table 4 Cont'd)

Element	Line	Wavelength	OH RI	OH LOD	AH RI	AH LOD	OA RI	OA LOD	AA RI	AA LOD
Thallium	Tl I	3775.72	10	0.1/B	10		10	0.09*	10	
		5350.46	7.0		6.0		5.0		7.0	
		3519.24	0.3		0.8		0.7		3.0	
		2767.87	0.05		-		0.5		1.0	
Thorium	Th	5760.55	NI		NI		10	150*	NI	
Thulium	TmO	5380	10	0.35/B	NI		NI		NI	
		5415	10							
		5530	10							
		4910	8.6							
		4830	6.3							
		5230	4.9							
	Tm I	4105.84	-		-		10	0.2*	-	
		4094.19	-		-		6.7	0.3*	-	
Tin	SnO	3585.4	10	25/A	7.0		1.8		1.0	
		3388.3	7.7		5.0		1.8		0.01	
		3415.8	7.7		6.0		0.71		0.3	
		3691.4	7.7		5.0		1.0		0.3	
		3721.2	7.7		5.0		0.89		0.3	
		3864.9	6.7		7.0		0.46		0.1	
		3833.2	5.6		5.0		0.46		0.3	
		3983.9	3.9		6.0		0.41		1.0	
		3323.5	2.8		3.0		1.8		1.0	
		3484.5	2.0		3.0		1.8			
		4850	1.0		10		-		-	
	Sn I	3262.34			0.8		2.4		10	

(Table 4 Cont'd)

			OH		AH		OA		AA	
			RI	LOD	RI	LOD	RI	LOD	RI	LOD
Tin Continued	Sn I	3034.12	0.55		0.8		2.4	4.0*	10	
		3009.14	0.44		0.4		1.8		7.0	
		2863.33	0.15		0.22		1.0		7.0	
		2839.99	(0.13)		0.16		1.2	20/A	10	
		2706.51	0.11		5.0		10		7.0	
		3801.02	-		0.24		1.8		7.0	
		3175.05	-		-		0.71		7.0	
		3330.62	-				2.4		5.0	
Titanium	TiO	5448	10		NI		1.5		NI	
		5759	10				1.5			
		5167	8.9				1.5			
		4955	7.8	0.2/A			1.1			
		5003	7.8				1.1			
		4805	6.7				0.7			
		4848	6.7				0.8			
		6730	4.4				10			
		7150	4.4				10			
	Ti I	3653.50	-				10	0.5*		
		3998.64	-				10	0.5*		
Tungsten	W	4008.75	NI		NI		10	4*	NI	
Uranium	(UO₂)	5500	10		10	0.1/A	-		-	
	U I	5915.40	-		-		10	10*	-	

(Table 4 Cont'd)

			OH		AH		OA		AA	
			RI	LOD	RI	LOD	RI	LOD	RI	LOD
Vanadium	VO	5469.3	10	0.3/A	9.1	0.02/B	4.0		0.1	
		5736.7	10		10		6.0		0.3	
		5228.3	7.5		7.3		3.0		0.01	
		5275.8	7.5		7.3		3.0		0.01	
		5058	6.25		6.4		1.2		-	
		6086.4	5.0		4.5		3.0		0.3	
		7100	2.50		-		10		7.0	
		8000	2.50		-		10		10	
		7470	1.75		-		7.0		10	
		5567.7	-		-		-		0.1	
	V I	4379.24	-		-		10	0.3*	-	
		4408.20	} -		} -		} 10	} 0.3*	} -	
		4408.51								
Xenon	Xe	No information								
Ytterbium	(YbOH)	5725	10	0.15/A	10	0.03/B	(10)	0.05*	7.0	
		5325	7.3		8.3		(5.0)		5.0	
		5550	} 6.4		} 8.3		(10)		7.0	
		5556.48								
	Yb I (YbOH)	4981	4.5		6.7		(5.0)		2.0	
		5174	3.6		5.0		(3.0)		0.1	
		5443	3.2		4.3		(3.0)		-	
	Yb I (YbOH)	3987.98	2.7		2.3		(10)		10	
		4850	2.7		3.3		(2.0)		1.0	
		5870	2.7		2.7		(2.0)		1.0	
		4778	2.0		3.3		(2.0)		2.0	
		6020	0.9		1.0		(1.0)		-	
	Yb I	3464.36	0.18		-		(0.5)		7.0	
	Yb II	3694.19	0.13						0.3	

(Table 4 Cont'd)

Element	Species	Wavelength	OH		AH		OA		AA	
			RI	LOD	RI	LOD	RI	LOD	RI	LOD
Yttrium	YO	5990	10	0.5/A	NI		10		10	
		6150	10				10		10	
		4818.2	1.0				–		7.0	
		4842.0	0.73				–		2.0	
		4676.3	0.33				–		2.0	
		5860	0.33				0.03		–	
		4706.7	0.30				–		1.0	
		4650.2	0.27				–		2.0	
		5746.9	0.27				0.03		0.01	
		5050	0.23				–		–	
		5078	0.20				–		–	
	Y I	4077.38	–				10	0.3*	–	
		4102.38	–				10	0.3*	–	
		4128.31	–				6.0	0.5*	–	
Zinc	(?) Zn I	5200-6000	10	0.8/A	10		–		NI	
		2138.56	5.0		0.8		10	50*		
		4810.53	5.0		6.0		–	1500*		
		3075.90	1.67		2.0		–			
Zirconium	ZrO	5640	10	8/A	NI		(10)		NI	
		5740	10				(10)			
	Zr I	3519.60	–				10	50*		
		3601.19	–				7.5	75*		

Table 4

FOOTNOTES

( )   means estimated values or emitting species

NI   means no information

1.   The symbol given is the emitting species and all lines given
     below the symbol are for the same species until another symbol
     appears, e.g., I refers to the atom line, II to the ion line,
     etc.

2.   Wavelengths conform to those given in NBS Monograph 32
     (Reference 42 in Section 1) where possible. If the line does
     not appear in the Monograph, the reference of R. Mavrodineanu
     is the second choice (Footnote 5 below), and the Handbook of
     Chemistry and Physics (Footnote 4) is third choice.

3.   Key to Flame Types.

             OH   -   oxygen-hydrogen flame
             AH   -   air-hydrogen flame
             OA   -   oxygen-acetylene flame
             AA   -   air-acetylene flame

4.   The data for the relative intensities for the oxygen-hydrogen,
     air-hydrogen, and oxygen-acetylene flames were compiled from
     the 48th edition of the Handbook of Chemistry and Physics,
     1967-68, Chemical Rubber Company, R. C. Weast, Ed., pp. E 136ff,
     except where noted for the oxy-acetylene flame (see footnote 5).

5.   Where noted *, the data for the oxygen-acetylene flame was
     taken from V. A. Fassel and D. W. Golightly, Anal. Chem., 39,
     466 (1967). These authors used a pre-mixed oxygen-acetylene
     flame which gives considerably increased sensitivities for
     many elements.

Data noted ** was taken from R. Mavrodineanu and Henri Boiteux, "Flame Spectroscopy," John Wiley & Sons, 1965.

6. All data for the air-acetylene flame was taken from R. Mavrodineanu and Henri Boiteux's reference cited above.

7. RI stands for relative intensity and is based on a scale where the most intense emission for a particular flame and an element is 10.

LOD stands for limit of detection and has units of parts per million. The LOD data was not generally from the same reference as the intensity data and is therefore only to be considered as representative. The letter after the LOD is the reference from which the number was taken (see below).

A. P. T. Gilbert, Jr., Beckman Bulletin 753, 1959.

B. P. T. Gilbert, Jr., Symposium on Spectroscopy, ASTM Spec. Tech. Publ., 1960.

C. P. T. Gilbert, Jr., Xth Colloquium Spectrocopicum Internationale, Spartan Press, 1963, Washington, D. C., pp. 171ff.

D. M. Honman, Anal. Chem., 27, 1656 (1955).

8. The data in Mavrodineanu's book is not given in a manner suited to comparison with photometric instruments; therefore, arbitrary numbers were applied according to the following designations in his reference:

very strong	10
strong	7
medium strong	5
strongly medium	3
medium	2
weakly medium	1
weak	.3
very weak	.1
very, very weak	.01

SECTION III

ATOMIC FLUORESCENCE SPECTRAL LINES

Section III

ATLAS OF ATOMIC FLUORESCENCE SPECTRAL LINES

Introduction

The inclusion of this section was also felt necessary for the sake of completeness. Atomic fluorescence spectroscopy is the youngest of the flame spectroscopic techniques. It originated in 1964 in Dr. Winefordner's group at the University of Florida. Therefore, relatively little time has been permitted for the complete investigation of very many elements. In fact, this compilation gives more information as to the effects of various flame types on the most sensitive lines for many of the elements rather than a study of the relative intensities of the various lines within the element. A few elements, however, have been studied surprisingly well in terms of the various fluorescing lines. This atlas points to the vast amount of research there is to be done in the area of atomic fluorescence spectroscopy.

Procedure

The procedure for this atlas is essentially the same as for the atomic absorption section. The most intensely fluorescing line in each reference is given a value of 10 and the less intense lines proportional values. As in the other sections of this atlas, the relative intensities should be taken only as a rule of thumb because the specific instrumental conditions used as well as the specific monochromator may well cause variations. The electronics package could also have a large effect on the relative intensities of the various lines.

Explanation of the Atlas

Again the atlas is relatively straightforward and requires little explanation. The elements are listed in alphabetical order by name. No differentiation of the symbols is necessary as in all

cases the transitions are atomic line transitions. The wavelengths
correspond to the NBS monograph (Reference 42, Section I), as in
the other sections. The relative intensities have already been
referred to. It should be noted that there are many more flame
types which have been used in atomic fluorescence spectroscopy than
in other atomic spectral methods; the key to these is given in
Table 6. In this section the limits of detection correspond to the
reference cited in the manner of Section I. Detection limits are
generally defined as a signal-to-noise level of 2. Finally, the
reference is given in the last column.

Discussion

Data for some 33 elements are presented in this section. Of
the 33 only 12 have been studied to any extent with respect to the
minor fluorescing lines. Special notice should be taken of the
variation of limit of detection as a function of flame type and/or
excitation source type. As mentioned before, considerable work
needs to be done in the area of atomic fluorescence, and it seems
certain that it will rapidly become possible to expand this section
of the atlas considerably.

TABLE 5

ATOMIC FLUORESCENCE LINES*

Element	Symbol	Line[1]	Relative Intensity	Flame[2] Type	Excitation[3] Source	Limit of[4] Detection	Reference
Actinium	Ac	No Information					
Aluminium	Al	3092.71⎫ 3092.84⎭	10	SAc/N	EDT	3	1
		3961.53	10	H/N	EDT	2	2
Americium	Am	No Information					
Antimony	Sb	2311.47	10	H/A	EDT	0.05	3
		2068.33	5				
		2598.05⎫ 2598.09⎭	5				
		2877.92	3				
		2769.95	2				
		2670.64	0.9				
		2510.54	NFO				
		2175.81	10	H/N$_2$	EDT	0.08	2
Argon	Ar	No Information					
Arsenic	As	2349.84	10	H/Ar	EDT	0.25	4,5
		2898.71	4				
		2492.91	4				
		2456.53	3				
		2860.44	3				
		2381.18	3				
		1971.97	3				
		1936.96	2				
		1890	2				
		2437.23	0.8				
		3032.85	0.7				
		2288.12	0.2				
		2730.22	0.04				
		2745.00	0.02				

*Footnotes for this table can be found on page 91.

(Table 5 Cont'd)

Arsenic Continued	As	2349.84	10	H/N$_2$	EDT	0.5	4,5
		2492.91	5				
		2381.18	4				
		2456.53	4				
		1971.97	3				
		2860.44	3				
		1936.96	3				
		1890	3				
		2437.23	1				
		3032.85	0.9				
		2898.71	0.4				
		2288.12	0.3				
		2780.22	0.04				
		2745.00	<0.04				
		2349.84	10	Ac/A	EDT	2	4,5
		2288.12	3				
		2456.53	3				
		2492.91	3				
		1936.96	2				
		1971.97	2				
		2381.18	2				
		1890	2				
		2437.23	0.5				
Astatine	At	No Information					
Barium	Ba	No Information					
Berkelium	Bk	No Information					
Beryllium	Be	2348.61	10	PAc/N	EDT	0.04	6
		2348.61	10	SAc/N	EDT	0.01	7
		2348.61	10	Ac/N	HIHC	0.5	8
		2348.61	10	Ac/O	HIHC	10	8
Bismuth	Bi	3024.64	10	H/Ar	EDT	0.5	9
		3067.72	5				
		2061.70	1				
		2696.76	0.6				
		2897.98	0.5				
		4722.19					
		4722.55	0.3				
		4722.83					
		2938.30	0.3				
		2230.61	0.3				

(Table 5 Cont'd)

Bismuth Continued	Bi	2989.03	0.3				
		2993.34	0.3				
		4121.53	0.1				
		2228.25	0.1				
		3397.21	0.1				
		3510.85	0.07				
		2627.91	0.07				
		2276.58	0.05				
		2780.52	0.04				
		3024.64	10	H/N$_2$	EDT	1	9
		3067.72	6				
		2061.70	1				
		2696.76	0.6				
		2897.98	0.4				
		4722.19 4722.55 4722.83	0.3				
		2230.61	0.2				
		2989.03	0.2				
		2938.30	0.2				
		2228.25	0.1				
		4121.53	0.07				
		3397.21	0.07				
		2627.91	0.04				
		3510.85	0.04				
		2276.58	0.03				
		2780.52	0.03				
		3067.72	10	H/A	EDT	0.005	1
Boron	B	No Information					
Bromine	Br	No Information					
Cadmium	Cd	2288.02	10	SAc/A	MVL	0.0005	10
		3261.06	0.005				
		2288.02	10	H/Ar	Xe9	0.08	11
		2288.02	10	P/A	Xe1.5	0.2	12
		2288.02	10	H/A/Ar	EDT	NI	13
		2288.02	10	H/A/N$_2$			
		2288.02	10	PH/A	EDT	NI	6
		2288.02	10	PH/Ar			
		2288.02	10	PH/N			
		2288.02	10	PH/O			
		2288.02	10	PAc/N			

(Table 5 Cont'd)

Calcium	Ca	4226.73	10	H/A	EDT	0.02	14
		4226.73	10	H/EA	Xe9	0.02	15
		4226.73	10	Ac/O	Xe9	NI	16
		No Fluorescence Observed		{H/O H/Ar			
Californium	Cf	No Information					
Carbon	C	No Information					
Cerium	Ce	No Information					
Cesium	Cs	8521.10	10	H/O/N_2	MVL	NI	17
Chlorine	Cl	No Information					
Chromium	Cr	3578.69	10	H/N_2	EDT	0.05	2
		3578.69	10	H/A	DHC	1	18
		3593.49	10	H/A	MVL	5	19
Cobalt	Co	2407.25	10	P/A	EDT	0.005	20
		2407.25	10	H/A		0.005	
		2407.25	10	Ac/A		0.01	
		2407.25	10	PH/A	HIHC	0.02	21
		2424.93	5				
		2521.36	4				
		2411.62	4				
		2432.21	2				
		2414.46 2415.30}	2				
		2528.97	1				
		2407.25	10	PH/O/Ar	HIHC	0.01	21
		2407.25	10	PAc/A		0.05	
		2407.25	10	H/O/Ar		0.01	
		2407.25	10	H/N_2	EDT	0.04	2
		2407.25	10	H/Ar/EA	Xe9	NI	16
		2407.25	10	H/O	EDT	2	22
		2407.25	10	H/Ar	EDT	0.1	14

(Table 5 Cont'd)

Copper	Cu	3247.54	10	H/A	HIHC	0.001	23
		3247.54	10	H/Ar	EDT	0.005	24
		3247.54	10	H/O	HIHC	0.1	8
		3247.54	10	H/EA		0.5	
		3247.54	10	H/N$_2$	EDT	0.03	2
		3247.54	10	P/A	Xe1.5	1	12
Curium	Cm	No Information					
Dysprosium	Dy	No Information					
Einsteinium	Es	No Information					
Erbium	Er	No Information					
Europium	Eu	No Information					
Fermium	Fm	No Information					
Fluorine	F	No Information					
Francium	Fr	No Information					
Gadolinium	Gd	No Information					
Gallium	Ga	4172.06	10	H/A	EDT	0.007	1
		4172.06	10	H/O	EDT	1	22
		4172.06	10	H/N$_2$	EDT	0.3	2
		4032.98	NI	H/A	MVL	NI	25
		4172.06	NI	H/A			
		4172.06	10	H/A/Ar	EDT	NI	13
		4172.06	10	H/A/N$_2$			
		4172.06	10	PH/A	EDT	NI	6
		4172.06	10	PH/Ar			
		4172.06	10	PH/N			
		4172.06	10	PH/O			

(Table 5 Cont'd)

Germanium	Ge	2651.18 2651.58	10	Ac/O/N$_2$	EDT	15	26
		2592.54	NI				
		2691.34	NI				
		2709.63	NI				
		2754.59	NI				
		2651.18 2651.58	10	Ac/N	EDT	3	1
Gold	Au	2675.95	10	H/A	DHC	0.05	27
		2675.95	10	H/O	EDT	0.2	21
		2675.95	10	H/Ar	Xe1.5	4	11
		2675.95	10	Ac/O	Xe9	NI	16
Hafnium	Hf	No Information					
Helium	He	No Information					
Holmium	Ho	No Information					
Hydrogen	H	No Information					
Indium	In	4511.31	NI	H/A	MVL	NI	25
		4104.76	NI				
		4104.76	10	H/Ar	EDT	0.1	21
		4511.31	10	H/N$_2$	EDT	1	2
Iodine	I	No Information					
Iridium	Ir	No Information					
Iron	Fe	2483.27	10	H/A	HIHC	0.02	21
		2522.85	2				
		2719.02	0.5				
		3020.49 3020.64	0.4				
		2483.27	10	H/O/Ar		0.02	

(Table 5 Cont'd)

Iron Continued	Fe	2483.27	10	Ac/A	HIHC	0.08	21
		2483.27	10	H/O/A		0.02	
		2483.27	10	H/Ar	EDT	0.3	24
		2483.27	10	H/N$_2$	EDT	0.008	2
		3734.87	0.2				
		2483.27	10	H/EA	Xe9	2	15
		2483.27	10	P/A	Xe1.5	5	12
		2483.27	10	H/A/Ar	EDT	NI	13
		2483.27	10	H/A/N$_2$			
		2483.27	10	PH/Ar	EDT	NI	6
		2483.27	10	PH/N			
		2483.27	10	PH/O			
		2483.27	10	H/O	Xe9	NI	16
		2483.27	10	Ac/O			
Krypton	Kr	No Information					
Lanthanum	La	No Information					
Lead	Pb	2833.06	10	H/N$_2$	EDT	0.06	2
		4057.83	2				
		4057.83	10	H/A	HIHC	0.02	23
		4057.83	10	H/O/Ar	Xe9	0.5	24
		4057.83	10	H/Ar		0.2	
		2833.06	10	P/A	Xe1.5	20	12
		2833.06	10	H/O	Xe9	NI	16
		2833.06	10	Ac/O			
Lithium	Li	6707.84	10	H/O/Ar	MVL	NI	28
Lutetium	Lu	No Information					
Magnesium	Mg	2852.13	10	P/A	HIHC	0.001	29
		2852.13	10	Ac/A		0.001	
		2852.13	10	Ac/N		0.005	
		2852.13	10	H/A	Xe9	0.002	1
		2852.13	10	H/Ar	Xe4.5	0.004	15

(Table 5 Cont'd)

Magnesium Continued	Mg	2852.13	10	H/EA	Xe4.5	0.01	30
		2852.13	10	H/O	EDT	0.01	22
		2852.13	10	H/O/Ar	Xe9	0.2	24
		2852.13	10	PH/A	EDT	NI	6
		2852.13	10	PH/Ar			
		2852.13	10	PH/N			
		2852.13	10	PH/O			
		2852.13	10	Ac/O	Xe9	NI	16
Manganese	Mn	2794.82 2798.27 2801.06	10	H/A	Xe4.5	0.003	1
		2794.82 2798.27 2801.06	10	H/Ar	EDT	0.006	21
		2794.82 2798.27 2801.06	10	H/N_2	EDT	0.01	2
		2794.82 2798.27 2801.06	10	H/EA	Xe9	0.1	15
		2794.82 2798.27 2801.06	10	P/A	Xe1.5	0.3	12
		2794.82 2798.27 2801.06	10	H/O	EDT	1	22
		2794.82 2798.27 2801.06	10	Ac/O	Xe9	NI	16
Mendelevium	Md	No Information					
Mercury	Hg	2536.52	10	H/A	MVL	0.1	31
		2536.52	10	H/O		0.1	
		2536.52	10	Ac/O		0.1	
		2536.52	10	NG/A		7	
		2536.52	10	H/N_2	EDT	0.1	2

(Table 5 Cont'd)

Molybdenum	Mo	No Information					
Neodymium	Nd	No Information					
Neon	Ne	No Information					
Neptunium	Np	No Information					
Nickel	Ni	2320.03	10	H/A	HIHC	0.005	21
		2310.96	0.5				
		3414.76	0.3				
		2289.98	0.2				
		2345.54	0.2				
		3002.49	0.2				
		3524.54	0.2				
		3050.82	0.1				
		3461.65	0.1				
		2312.34	<.1				
		2313.98	<.1				
		2325.79	<.1				
		3003.63	<.1				
		3101.55⎫ 3101.88⎭	<.1				
		3134.11	<.1				
		3515.05	<.1				
		2320.03	10	H/O/Ar		0.003	
		2320.03	10	Ac/A		0.03	
		2320.03	10	H/O/Ar		0.006	
		2320.03	10	H/EA	Xe9	1	15
		2320.03	10	H/N$_2$	EDT	0.006	2
		3414.76	0.6				
Niobium	Nb	No Information					
Nitrogen	N	No Information					
Nobelium	No	No Information					
Osmium	Os	No Information					
Oxygen	O	No Information					

(Table 5 Cont'd)

Palladium	Pd	3404.58	10	H/A	SHC	2	32
Phosphorus	P	No Information					
Platinum	Pt	No Information					
Plutonium	Pu	No Information					
Polonium	Po	No Information					
Potassium	K	7664.91 7698.98	NI NI	H/O/Ar	MVL	NI	28
Praseodymium	Pr	No Information					
Promethium	Pm	No Information					
Protactinium	Pa	No Information					
Radium	Ra	No Information					
Radon	Rn	No Information					
Rhenium	Re	No Information					
Rhodium	Rh	No Information					
Rubidium	Rb	7947.60	10	H/O/Ar	MVL	NI	28
Ruthenium	Ru	No Information					
Samarium	Sm	No Information					
Scandium	Sc	No Information					
Selenium	Se	2039.85 1960.26 2062.79 2074.79	10 7 5 0.2	P/A	EDT	0.2	33

(Table 5 Cont'd)

Selenium Continued	Se	1960.26	10	H/Ar	EDT	0.4	14
		1960.26	10	H/O		0.8	
		1960.26	10	H/N$_2$		0.2	
		2039.85	10	H/N$_2$	EDT	0.2	2
		1960.26	10	H/A		0.04	
Silicon	Si	2516.11	10	Ac/N	EDT	2.	34
		2506.90	NI				
		2514.32	NI				
		2519.21	NI				
		2524.11	NI				
		2528.51	NI				
		2516.11	10	SAc/N	EDT	0.5	34
Silver	Ag	3382.89	10	P/A	Xe1.5	0.2	12
		3280.68	6				
		3382.89	10	H/A		0.1	
		3280.68	10				
		3280.68	10	H/O	Xe4.5	0.003	30
		3280.68	10	H/EA		0.001	
		3280.68	10	H/Ar	EDT	0.0005	24
		3280.68	10	H/N$_2$	EDT	0.1	2
		3280.68	10	H/O/Ar	Xe9	0.5	24
		3280.68	10	Ac/A	HIHC	0.004	35
		3280.68	10	Ac/O	Xe9	NI	16
Sodium	Na	5895.92	10	H/A	Xe4.5	0.008	1
		5895.92	NI	H/O/Ar	MVL	NI	17
		5889.95	NI				
Strontium	Sr	No Information					
Sulfur	S	No Information					
Tantalum	Ta	No Information					
Technetium	Tc	No Information					

(Table 5 Cont'd)

Tellurium	Te	2383.25⟩ 2385.76⟨ 2142.75 2259.04 2530.70	10 7 0.4 NFO	P/A	EDT	0.05	33
		2142.75	10	H/O	EDT	0.5	22
		2142.75	10	H/Ar	EDT	0.5	14
		2142.75	10	H/A	EDT	0.006	1
		2142.75	10	H/N$_2$	EDT	0.06	2
Terbium	Tb	No Information					
Thallium	Tl	3775.72 5350.46 2767.87	10 4 1	H/A	EDT	0.1	36
		3775.72	10	H/Ar	EDT	0.008	14
		3775.72 3775.72	10 10	H/A/Ar H/A/N$_2$	EDT	NI	13
		3775.72 3775.72 3775.72 3775.72	10 10 10 10	PH/A PH/Ar PH/N PH/O	EDT	NI	6
		3775.72	10	H/O	MVL	0.04	37
		3775.72	10	H/N$_2$	EDT	0.2	2
		3775.72 3775.72	10 10	Ac/O NG/A	MVL	5 1	31
		3775.72	10	H/EA	Xe4.5	0.07	15
		3775.72	10	H/O/Ar	MVL	NI	38
Thorium	Th	No Information					
Thulium	Tm	No Information					

(Table 5 Cont'd)

Tin	Sn	3034.12	10	H/O/Ar	EDT	0.1	39
		3175.05	9				
		2863.33	5				
		2839.99	4				
		3009.14	4				
		2706.51	2				
		3801.02	1				
		3330.62	0.7				
		3262.34	0.2				
		2546.55	0.1				
		2429.49	0.1				
		2246.05	0.07				
		2354.84	0.07				
		3175.05	10	H/Ar	EDT	0.3	39
		3034.12	6				
		2863.33	6				
		3009.14	4				
		2839.99	3				
		2706.51	3				
		3801.02	2				
		3262.34	2				
		2429.49	0.3				
		2546.55	0.2				
		3330.62	0.2				
		3034.12	10	Ac/A	EDT	0.6	39
		2839.99	8				
		3175.05	7				
		2863.33	4				
		3009.14	3				
		2706.51	3				
		3262.34	2				
		3801.02	1				
		3330.62	0.9				
		2429.49	0.4				
		2354.84	0.3				
		2546.55	0.3				
		2246.05	0.2				
		3034.12	10	SH/O/Ar	EDT	0.1	39
		3034.12	10	H/A	EDT	NI	13
		3034.12	10	H/A/Ar			
		3034.12	10	H/A/N_2			
		3034.12	10	H/N_2	EDT	0.6	2
Titanium	Ti	No Information					
Tungsten	W	No Information					

(Table 5 Cont'd)

Uranium	U	No Information					
Vanadium	V	No Information					
Xenon	Xe	No Information					
Ytterbium	Yb	No Information					
Yttrium	Y	No Information					
Zinc	Zn	2138.56	10	H/A	EDT	0.00004	40
		2138.56	10	H/O	MVL	0.0002	37
		2138.56	10	H/N$_2$	EDT	0.0002	2
		2138.56 2138.56	10 10	Ac/O NG/A	MVL	0.04 0.01	31
		2138.56	10	H/EA	Xe4.5	0.01	15
		2138.56	10	P/A	MVL	0.003	12
		2138.56	10	SAc/A	MVL	0.0002	10
		2138.56	10	Ac/A	MVL	NI	41
		2138.56 2138.56 2138.56 2138.56	10 10 10 10	PH/Ar PH/N PH/O PH/A	EDT	NI	15
Zirconium	Zr	No Information					

Table 5

FOOTNOTES

1. Wavelengths conform to NBS Monograph 32 where possible.

2. See Table 6 (p. 92) for Flame Type code.

3. See Table 7 (p. 93) for Excitation Source code.

4. The limit of detection values are generally defined as a signal-to-noise ratio equal to 2 and is given in parts-per-million.

   NI means no information

   NFO means no fluorescence observed at the indicated wavelength.

Table 6

ATOMIC FLUORESCENCE FLAME TYPE CODE

H/O	Oxygen-hydrogen
PH/O	Pre-mixed oxygen-hydrogen
H/A	Air-hydrogen
PH/A	Pre-mixed air-hydrogen
H/Ar	Hydrogen-argon-entrained air
PH/Ar	Pre-mixed hydrogen-argon-entrained air
H/O/Ar	Hydrogen-oxygen-argon
H/O/N_2	Hydrogen-oxygen-nitrogen
H/N	Hydrogen-nitrous oxide
PH/N	Premixed hydrogen-nitrous oxide
H/EA	Hydrogen-entrained air
H/N_2	Hydrogen-nitrogen entrained air
P/A	Propane-air
Ac/A	Acetylene-air
SAc/A	Separated acetylene-air
Ac/O	Acetylene-oxygen
Ac/N	Acetylene-nitrous oxide
SAc/N	Separated acetylene-nitrous oxide
PAc/N	Premixed acetylene-nitrous oxide
NG/A	Natural gas-air
Ac/O/N_2	Acetylene-oxygen-nitrogen
SH/O/Ar	Separated hydrogen-oxygen-argon

Table 7

ATOMIC FLUORESCENCE EXCITATION SOURCE CODE

EDT	Electrodeless discharge tube
HIHC	High intensity hollow cathode discharge tube
MVL	Metal vapor discharge arc lamp
Xe1.5	150 Watt xenon arc lamp
Xe4.5	450 Watt xenon arc lamp
Xe9	900 Watt xenon arc lamp
DHC	Demountable hollow cathode discharge tube
SHC	Shielded hollow cathode discharge tube

Table 5

REFERENCES

1. A. Hell and S. Ricchio, Talk given at the Pittsburgh Confer-
    ence on Analytical Chemistry and Applied Spectroscopy,
    Cleveland, Ohio 1970. (Paper No. 23).

2. R. M. Dagnall, M. R. G. Taylor, and T. S. West, Spectr.
    Letters, 1, 397 (1968).

3. R. M. Dagnall, K. C. Thompson, and T. S. West, Talanta, 14,
    1151 (1967).

4. R. M. Dagnall, K. C. Thompson, and T. S. West, Talanta, 15,
    677 (1967).

5. R. M. Dagnall and T. S. West, Appl. Optics, 7, 1287 (1968).

6. M. P. Bratzel, R. M. Dagnall, and J. D. Winefordner, Anal.
    Chem., 41, 1527 (1969).

7. D. N. Hingle, G. F. Kirkbright, and T. S. West, Analyst, 93,
    522 (1968).

8. J. W. Robinson and C. J. Hsu, Anal. Chim. Acta, 43, 109
    (1968).

9. R. M. Dagnall, K. C. Thompson, and T. S. West, Talanta, 14,
    1467 (1967).

10. P. S. Hobbs, G. F. Kirkbright, M. Sargant, and T. S. West,
    Talanta, 15, 997 (1968).

11. C. Veillon, J. M. Mansfield, M. L. Parsons, and J. D.
    Winefordner, Anal. Chem., 204 (1966).

12. R. M. Dagnall, K. C. Thompson, and T. S. West, Anal. Chim. Acta, 36, 269 (1966).

13. M. P. Bratzel, R. M. Dagnall, and J. D. Winefordner, Anal. Chem., 41, 713 (1969).

14. K. E. Zachi, M. P. Bratzel, J. M. Mansfield, and J. D. Winefordner, Anal. Chem., 40, 1733 (1968).

15. D. W. Ellis and D. R. Demers, "Atomic Fluorescence Flame Spectrometry," in Trace Inorganics in Water, R. A. Bellar, ed., Advances in Chemistry Series, No. 73, Washington, D. C., 1968.

16. S. J. Pearce, L. deGalan, and J. D. Winefordner, Spectrochim. Acta, 23B, 793 (1968).

17. D. R. Jenkins, Proc. Roy. Soc., London, A 293, 493 (1966).

18. G. Rossi and N. Omenetto, Talanta, 16, 263 (1969).

19. N. Omerretto and G. Rossi, Anal. Chim. Acta, 40, 195 (1968).

20. B. Fleet, K. V. Liberty, and T. S. West, Anal. Chim. Acta, 45, 205 (1969).

21. J. Matousek and V. Sychra, Anal. Chem., 41, 518 (1969).

22. J. M. Mansfield, M. P. Bratzel, H. O. Norgardon, D. N. Knapp, K. E. Zacha, and J. D. Winefordner, Spectrochim. Acta, 23B, 389 (1968).

23. D. L. Manning and P. Heneage, At. Abs. Newsletter, 6, 124 (1967).

24. R. Smith, C. M. Stafford, and J. D. Winefordner, Can. Spectr., 14, 2 (1969).

25. N. Omenetto and G. Rossi, Spectrochim. Acta, 24B, 95 (1969).

26. R. M. Dagnall, K. C. Thompson, and T. S. West, Anal. Chim. Acta, 41, 55 (1968).